Language of Life

Peirce Studies
Peirce Studien
Studi Peirce
Études Peirce
Estudios Peirce

Number 12

Peirce Studies is a peer-r eviewed international monographic series of the Institute
for Studies in Pragmaticism at Texas Tech University, Lubbock, Texas, 79409-0 002,
USA, to which address all editorial matters should be sent. The preferred language
is English; however, publication in the other international languages may occur.

Series Editor
Elize Bisanz
Editorial Board
Kenneth L. Ketner
Lisa Palafox
Rhonda McDonnell
Michael O'Boyle
Karey Perkins
Stephanie Schneider

Ľudmila Lacková

Language of Life
A Peircean Approach to Living Organisms

PETER LANG

Berlin - Bruxelles - Chennai - Lausanne - New York - Oxford

Bibliographic Information published by the Deutsche Nationalbibliothek
The Deutsche Nationalbibliothek lists this publication in the
Deutsche Nationalbibliografie; detailed bibliographic data is
available in the internet at http:// dnb.d- nb.de.

Library of Congress Cataloging-in-Publication Data
A CIP catalog record for this book has been applied for at the Library of Congress.

This book publication was supported by an internal research grant from the
Faculty of Arts, Palacký University in Olomouc (2019-2021), *Aplikace sémiotických
popisů na rozhraní pragmatiky a sémiotiky*, project number FPVC2019/03.

All of the illustrations in the book are by Mia Lévyová.

ISBN 978-3-631-92593-5 (Print)
E- ISBN 978-3-631-92594-2 (E- Book)
E- ISBN 978-3-631-92595-9 (E- Pub)
DOI 10.3726/b22291

© 2025 Peter Lang Group AG, Lausanne, Switzerland

Published by Peter Lang GmbH, Berlin, Germany

info@peterlang.com www.peterlang.com

Table of Contents

List of Abbreviations

CP: Peirce, C.S. 1994 [1866–1913]. *The Collected Papers of Charles Sanders Peirce*. Hartshorne, Ch., Weiss, P. (eds.). 1931–1935. Electronic edition reproducing. Vols. I–VI. Cambridge: Harvard University Press. Vols. VII–VIII Burks, A. W. (ed.). 1958. Cambridge: Harvard University Press. Charlottesville: Intelex Corporation.

CD: The Century Dictionary and Cyclopedia *1889–1891*, Editor-in-Chief, William D. Whitney, New York: The Century Company, 10 volumes. Online version by the Institute for Studies in Pragmaticism archives, Texas Tech University, Lubbock: http://www.global-language.com/century/,

EG: Existential Graphs

EP1:   The Essential Peirce, Volume 1: Selected Philosophical Writings (1867–1893). (1992). United States: Indiana University Press.

I: Interpretant of semeiosis

ISP: Interdisciplinary Seminar on Peirce, Elize Bisanz, Kenneth Ketner, Clyde Hendrick, Levi Johnson, Michael O'Boyle, Thomas G. McLaughlin, Scott Cunningham

MS: Peirce manuscripts from Institute for Studies in Pragmaticism archives, Texas Tech University, Lubbock.

NEM: Peirce, Charles S., *The New Elements of Mathematics*, Vol. I–IV. Carolyn Eisele, ed. (De Gruyter, 1976)

NRT: NonReduction Theorem

O: Object of semeiosis

PS: Peirce Studies, peer-reviewed monographic series edited by members of the Institute for Studies in Pragmaticism at Texas Tech University

R: Representamen of semeiosis

SS: *Semiotics and Significs: Correspondence Between Charles S. Peirce and Victoria Lady Welby*, 2nd edition, in PS n. 8, Charles S. Hardwick (ed.), The Press of Arisbe Associates, Elsah, Illinois.

List of Manuscripts

MS 467, Lowell Lectures 1903, Lecture 4, Vol 1: Gamma Part of Existential Graphs
MS 1334, Adirondack Summer School Lectures, 1905
MS 478, Syllabus of a Course of Lectures at the Lowell Institute beginning, 1903
MS 870, What is a Law of Nature? 1894
MS 837, [Various Topics in Logic], not published, not dated
MS 482, On Logical Graphs, 1896

List of Figures

Chapter 1 Growth

Introduction

Understanding our biological nature is intertwined with a very intimate relation of each person to individual body and mind as well as with the definition of who we are as a species, what is our part in the biosphere and nature in general. The impact of biology on society has been immense in recent history, both in negative and positive ways, and it has raised many controversies related to worldviews, religions, and philosophical standpoints with enormous influence on general common-sense reasoning developed in the course of the last century. Charles Sanders Peirce perceived a direct connection between Darwinian evolution and the "gospel of greed" as he called the development of capitalist society. But this critical review of Darwin came only after decades of general acceptance of *On the Origin of Species* and the theory of random variation and survival of the fittest. At the time of its publication the controversies were more serious and critique was relentless. The major concern was related to a shift in deterministic explanations of evolution due to the notion of chance and randomness, and related worries by representatives of the church, but alongside the church, criticism came also from the scientific community[1]. Another milestone in modern biology, the decipherment of the genetic code, was perceived by society, on the contrary, with great welcome and general enthusiasm. Many scientists who contributed to the decipherment of the secret language of genes became winners or laureates of the Nobel prize. Reluctance towards Darwin's theory and the warm welcoming of the genetic code are very likely due to the same attitude towards deterministic explanations of life phenomena.

In the former case the motive was distance-taking from determinism, while in the latter case the theme was a restoration of the deterministic laws governing the smallest molecular process of living organisms, of *us*. The genetic script is publicly perceived as an ideal, flawless and very reassuring mechanism identifying precisely and unambiguously who we are. This fact had many implications for social institutions like jurisprudence, social law, criminal law, the health

---

1   In popular and religious thinking in the 19th century, Darwin's theory contradicted the idea of special creation and, much like the Copernican Revolution, threatened to displace human beings as the center of value in the cosmos. To suggest that chance variation was the source of our coming-into-being, was to infer that the cosmic order lacked intelligence and, hence, a divine architect. (Liszka 2014: 145)

system, parenthood and family law. The public understanding of the flawlessness of DNA analysis is somewhat romantic, naïve and idealized. It has been accepted as the very definition of life by the general public. Take a hair or a piece of tissue to the laboratory and you will know to whom it belongs. The DNA molecule appears to be a perfect tool for reducing criminality. Yet there have been countless cases where the attribution of the crime based on DNA evidence has been unclear or even erroneous. Remember the case of Yara Cambisario from Italy in 2010. The case caught media attention because the first DNA match pointed to a ten year-old boy, but after further analysis the actual culprit was identified as the illegitimate son of the first DNA match. There are still doubts and controversies over this case, only reinforcing what since the beginning was already apparent: DNA is not unequivocal or unambiguous.

In the last couple decades, the Modern Synthesis of Darwinism – which represents evolution based on the transmission of genes through replication from generation to generation – has been called into question by new findings in epigenetics and evolutionary developmental biology. The Extended Evolutionary Synthesis (EES) has started to compete with the Modern Synthesis. By the term "synthesis" in this terminology a synthesis of Darwin's work is meant. For decades, the Modern Synthesis of Darwinism represented the mainstream in evolutionary studies, picking up only selectively parts from Darwin's theory and completing these with new discoveries from genetics. The Modern Synthesis embraces genetic determinism and the dogma of the unidirectional transfer of hereditary information. EES is trying to replace genetic determinism with a more open approach, considering epigenetics and evolutionary developmental biology as fields where the transfer of hereditary information is not exclusively genetic and is not unidirectional. From this new perspective, the genetic code appears not to be absolutely deterministic, organisms are presented as agents rather than passive vehicles of genes or replicators, and genes have been weakened in their infinite immutability and all-governing power. So far, we cannot speak about a change in the paradigm; the Modern Synthesis retains its primary position as a powerful explanatory framework. Yet epigenetics is gaining more and more consideration due to increasing experimental evidence. Despite this, the reception by the scientific community has been far from enthusiastic, and as far as the general public is concerned, epigenetic discoveries have not even achieved comparable popularity.

The discovery of the enormously long strings of non-coding DNA is somehow less attractive than the discovery of the general principle of life hidden in *only four letters* of the genetic alphabet. It lacks the elegant minimalism of the genetic code formula and it only brings chaos to such a beautifully simple definition of life. Uncomfortable as it might be, the research is advancing, and new experimental facts demonstrate that molecular aspects of life are not so simple and elegant as has been long supposed. It is an ambitious and demanding task to theoretically embrace such a radical change in approach and provide boundaries for the far-reaching understanding of it, let alone its implications for the question of what it means to be alive, what *we are* in biological terms, and what are the consequences for culture and society. A theoretical framework is needed for encompassing these new facts about life, and it shall not be a deterministic framework, because hard determinism is inherently excluded by the very definition of epigenetics.[2]

Current advances in biology are not compatible with the popularized theoretical movements of passive genetic replicator machines anymore. A new theoretical approach will inevitably replace the current general image of genetic inheritance, which will propose a solid and scientifically acceptable explanation to variation in molecular processes leading to the expression of genes, the active role of the environment, organisms, and the dynamic process of inheritance. Without falling back on simple vitalism, pre-deterministic kinds of teleology, or other concepts incompatible with the current scientific paradigm, this new framework for biology will provide toolkits for not only describing biological phenomena, but also its transdisciplinary bridging. It is apparent that the first and most urgent change needed is to get rid of dyadic explanations based on binary oppositions such as genotype-phenotype, DNA-protein, individual-community, but also male-female, life-death, and finally mind-matter. One of the changes needed for thinking in biology has to do with the embrace of topology and diagrams in place of strictly linear models. The theory of scientific reasoning offered by Charles Sanders Peirce fulfils all the requirements for a suitable theoretical approach towards the newest trends in biology. In this book, a handful of the most illuminating applications of Peirce are freshly presented, with the hope that his logical tools for scientific reasoning might find a new place in the biology of the 21st century. From these logical tools we will implement mostly the Existential Graphs, and more concretely the version by ISP (2019) called Betagraphic, which is a modification of Peirce's original Beta Graphs. With the aid of Peircean tools we will study epigenetics, protein folding and other biological processes at the molecular level. We will also re-visit the language metaphor of life and compare Peirce's thinking with the structural school of linguistics. With the ambition of "decentering" (in the Derridean sense) both linguistics and

---

2   Later in the text the case of epigenetic determinism also will be mentioned.

biology from a dyadic linear approach, this book will build the bridge between the problematic humanities and natural science, both in method and in the object of study. C. S. Peirce is probably the only figure to follow in such endeavors. In the introduction to the tenth volume of the series *Peirce Studies*, the editor of the series Elize Bisanz characterized C. S. Peirce in this light.

> A leading figure of this movement [aiding to bridge the gap between Humanities and Natural Science][3], Charles S. Peirce, devoted most of his writings to the foundation of a logic which would embrace the study of all kinds of relations found in natural and human interaction. (Bisanz 2019: 9)

The diagrams and iconic logic of Peirce represent an overall method for this book. Less a tool for a proper logical analysis and more an invitation to free our minds from certain ways of thinking, this book is an invitation to the readers, scholars in different fields, to circumnavigate the dictate of the linearity of speech. Peirce's diagrams are invitation to embrace a new scientific method, to open up our minds and try the reasoning in graphs instead of reasoning in the prison of the linear. Whether we are dealing with genetic linear strings, peptide chains, or human speech transcriptions, Peirce always reminds us that these are only a few of the countless possible ways of representing the given phenomena. This book is an attempt to show to the reader that there are other ways for scientific research, other ways for reasoning, other ways for *writing*. By no means exhaustive, this book provides just a few examples of what direction future research in natural science and the humanities will take.

Semeiotic of Seeds

Dyadic and binary logic has been applied to biology since ancient times. Peirce's philosophy and logic arose in the context of natural philosophy, largely influenced by Aristotle. We find reference to Aristotle in practically every paper by Peirce on nature or biology. The Aristotelian influence was primarily binary. Let's begin with an analysis of the entry "nature" from the *Century Dictionary* (CD) of which Peirce is the author (as he is author of all the entries in the fields of Logic, Metaphysics, Mathematics, Mechanics, Astronomy, Weights and Measures, and Color Terms in this encyclopedia). In defining nature, the duality of opposite forces governing nature is depicted (meaning n. 3 of the entry "nature" in the *Century Dictionary*):

---

3   Inserted by the author of this book.

The metaphysical principle of life; the power of growth; that which causes organisms to develop each in its predetermine way. Aristotle defines nature as the principle of motion in those things that move themselves, meaning by motion especially generation and corruption. Inasmuch as the most striking characteristic of growth is its regularity, nature is also conceived by Aristotle as the principle of inward necessity, as opposed to constraint on the one hand and to chance or freedom on the other. Hence *nature* is in literature frequently contrasted with fate and with *compulsion*, as well as with *fortune* and *free election*.[4] (CD: 3943, accessed 25/2/2024)

Peirce continues in this entry referring to Aristotle and his definition of nature as the constant tendency aspiring to "perpetual renovation of forms as perfect as they may be", but disturbed by the variable Aristotle calls "spontaneity and chance, forming an independent agency inseparably accompanying nature – always modifying, distorting, frustrating the full process of nature" (Grote, Aristotle: IV). The duality of a constant force and a variable force is an inseparable whole defining nature. The two forces are accordingly represented by the tendency to grow into perfect forms on the one hand, and randomness, chance, and pure spontaneity on the other hand. It will be demonstrated later in this book that the two opposite forms acquire in Peirce's own writings the names of law and chance. The topic is elaborated in a paper from 1894 by Peirce titled *Law of Nature* (MS 870)[5], where the author attributes the notion of law to the process of biological growth. The law of nature explains the process of growth in living organisms, the age-old mystery of the growth of differentiated and formed organisms from formless seeds in plants or eggs in animals. In our days, the mystery is seemingly resolved[6]; we would say, growth is encoded in the genetic information of the germ cell. Yet "genetic information" and "encoded" are only words, and no one in reality has a concrete definition of what they mean. In recent decades the very notion of gene is being questioned (Jost 2007): what is a gene? Are genes to be defined in terms of strings or in terms of the corresponding biological functions they encode. If we decide on the second option, with new advances in genetics and epigenetics we must admit that there is no such straight and unidirectional relation between genes and the function they encode. Moreover, many of the biological functions are not even present in the DNA

---

4   Italics are in the entry by Peirce and the difference in the font size is taken from the *Century Dictionary* layout.

5   Manuscripts cited in this book are housed at the Institute for Studies in Pragmaticism, Texas Tech University, Lubbock.

6   It has to be noted here that, nevertheless, ontogenesis is not fully explained by genetic causality.

information; they only emerge in later states of the protein synthesis and with contact of the environment. Thus, dealing with recent doubts and reconsiderations of well-established biological notions like "gene", we are in a way returning to an age-old mystery. Seeds and eggs were replaced by genes, yet this is only an insignificant difference related to the focus on the even more microscopic, *molecular* level achieved through progress in technology; but this technological progress only amplifies what was already present in ancient times: the astonishment and fascination with the paradox between the size of the minuscule particle (germ cells/genes) and its immense potentiality to grow. In both the cases of seeds and genes the shared characteristics are the minuscule dimensions and pluripotent, almost omnipotent power they hide: a huge and diversified organism grows from a tiny particle. The question *what is growth* is re-opened in biology. In this light, Peirce's manuscript MS 870 gains new relevance. Today's biology confronts the very same problem of the astonishing development of various biological forms from a pluripotent minuscule seed.

There is a connection between the concept of seed and Peirce's theory of signs, or better yet, Peirce's theory of semeioses. Before I proceed to the explanation of this connection, it is crucial to determine some basic notions from Peirce's semeiotic. My usage of the semeiotic terms in this book is loyal to the terms as defined and used by Institute for Studies in Pragmatism at Texas Tech University:

> Sometimes the discussion of semeiotic is undertaken as a study of the "theory of signs". This approach has produced some useful studies, however, there is an ambiguity problem that should be resolved. In that context the word "sign" can mean (1) the item (representamen) that represents the object to an interpretant within a semeiosis relation, or it can mean (2) the entire triadic semeiosis relation involving an object, a representamen, and an interpretant. Sense (1) is the narrow meaning of "sign"; whereas sense (2) is a broad sense of the word. These two senses are seriously different. Clarity is important in objective study, so we will drop "sign" and use "semeiosis" for the broad meaning (2) and use "representamen" for the narrow sense (1). (Ketner 2023: 11)

Peirce's semeiotic theory presupposes never ending, unlimited interpretive chains – in other words, semeiosis has no beginning and no end, exactly like life has no beginning and no end. It was not accidental that Peirce came to semeiotic through years long study and academic work in natural sciences. Semeiosis does not start with a single representamen and its interpretation. Every representamen is only another interpretation of a previous representamen. Likewise, life does not start with a seed. A seed is already the result of the biological functions of another organism and leads to further life-creation and procreation. Etymologically, the word semeiosis derives from Greek *seme* (seed), so it is already apparent from the very word that semeiotic is concerned with growth

(Ketner 2011: 379). Thanks to the close relation between semeiotic and the notion of growth, we can conclude a certain universality of growth extending beyond natural science. Semeiotic is an interdisciplinary framework, thus by extension also the notion of growth may become interdisciplinary. Not by metaphor, but by a real, ever-growing principle in nature and culture, these can be generalized under the term semeiotic growth. We can understand the notion of growth in terms of increasing complexity and differentiation, but also in terms of habit-making and habit-breaking and the consequent creation of new habits. In this understanding, semeiosis is living, thoughts are living, and growth is the generalizing principle unifying nature and culture.

At this place we can focus on MS 870 on the very term *law* of nature: growth is defined as the result of the law of nature. The law is, in simple words, the general tendency to grow to a given form from an original germ cell. We should not get confused by the term *law*, implying growth as the deterministic becoming of pre-given forms, structures, or preconceived ideal morphologies. Even though Peirce starts this manuscript with Aristotle, in the course of the text he takes distance and presents a critical standpoint toward Aristotle. The strict dualism between the lawlike constant of the tendency to grow into predetermined forms and random variable of chance is broken or attenuated by Peirce in his evolutionary theory, and the introduction of the notion of habit as a third element completes the dichotomy of law and chance, creating a trichotomy, a triadic evolution. In MS 870 this trichotomy is not introduced, but it is clear that Peirce's understanding of law is different from that of Aristotle. Peirce's law of nature is closer to habit. As will be demonstrated in the next chapter, tension and oscillation between law and habit is inherent to a very thin frontier between anancasm and agapasm.

In addition to the duality of chance (variable) and law (constant) in Aristotle, Peirce mentions another ancient lifegiving duality: the male and female element in nature. Growth is the lifegiving principle (Peirce, *The Basis of Pragmaticism*, in Bisanz 2009, 261) in life, culture and science. Male and female forces are also by extension the elementary principles for growth beyond biology, representing opposite polarities and the force of attraction between them (of course, I am excluding phenomena such as parthenogenesis, for the sake of simplicity). A similar principle is present in physics and other natural sciences, yet is also present in culture. In living nature, the female function is the function of the seed and female is the general and essential sex, according to Peirce: "the male only executes the hunch" (Peirce, *The Basis of Pragmaticism*, in Bisanz 2009, 262). Peirce argues that pure feminity as a quality can be conceived but cannot be realized while pure masculinity cannot be even conceived, it is "absurdity and nonsense". These two prerequisites for life, female and male forces, are not sufficient for the emergence of life. They must be completed by something third thanks to which growth is possible. Peirce does not specify what this Third is in *Basis of Pragmaticism*, but he specifies that this need of the third element for growth is applied to knowledge as well.

In another text (MS 837), the term of evolution is presented as a particular instantiation of the principle of growth. Evolution of species is a specific kind of growth, where it is not single organisms that grow (ontogeny), but whole lineages (phylogeny). This kind of growth shares its structural features with the growth of single organism development; that is, it proceeds from uniformity to diversity, differentiation, increase in complexity. In this manuscript, Peirce treats the notion of evolution in an unusual way. Its meaning is widened beyond biology (evolution of species). The reader of the manuscript might get confused because the text starts with introducing evolution and the names of Darwin and Lamarck, but then all remaining pages are a focused exposition of the development of the field of formal logic. In fact, the manuscript is a study of the changes in logical thinking across centuries, starting again with Aristotle and continuing towards the present day of Peirce. Logic as a concrete part of science being itself a concrete part of human creations, follows the law of evolution, as every human creation follows the law of evolution in this extended sense. Philosophy (natural philosophy) follows this law even more distinctly. In the history of the development of Aristotle's syllogism many changes were made. Peirce is especially interested in the reciprocal aspect of logical relations in the context of De Morgan's insights. De Morgan brought to logic, besides negations, the principles of identity and transitive relations. Transitive relations, which allow for reciprocity, were not described by Aristotle even though they are so important in classical syllogisms.

For instance, the relation of LOVE is not a transitive relation: it does not require reciprocity for the meaning of the word to remain unchanged. If A loves B, this does not imply that B loves A (in an ideal romantic way of course, reciprocity is desired, but in logical analysis this aspect is not taken into account). This is different from the principle of identity, where if A is B then it is also true that B is A (or part of it). All these observations develop the beautifully simple classical syllogism to an extremely complex form. In this manner, the tendency to growth with accompanying increase in complexity and differentiation is fulfilled. In The Basis of Pragmaticism (Peirce in Bisanz 2009, 258–286)) the reciprocal relations are described by Peirce as governing principles in reasoning in general.

> Reciprocal action,—action and reaction,—give duality, however slight the energy of the action may be. For here [are] two objects which belong together in the sense of being such as they are only as acting upon, and being acted upon by, each other. All inhibition of action, or action upon action, involves reaction and duality. All self-control involves,

and chiefly consists in, inhibition. All direction toward an end or good supposes self-control, and thus the normative sciences are thoroughly infused with duality. (Peirce, The Basis of Pragmaticism, in Bisanz 2009, 273)

Duality here is meant in the sense of reciprocal action and reaction in the activity of reasoning as only a sub-category of growth. Growth and self-control are two principles with similar difficulties, while they are necessary parts of any sign system, consequently of consciousness: a sign system with self-control leads to consciousness, according to Peirce. He talks about, for example, the consciousness of a fish, as already a result of growth and self-control (Peirce, The Basis of Pragmaticism, in Bisanz 2009, 276).

The main message of this excursion to the "evolution of logic" is that *reasoning evolves*, but Peirce asks: what kind of evolution is this? Darwinian or Lamarckian?

> Logic, therefore, tends to back up the suggestion of biology that the whole observation ought to be regarded as an evolution. But if that is once assumed, the question arises what is the method of this evolution. Is it Darwinian, in the main, or is it Lamarckian or what is it? Now if the law of evolution governs every domain of the universe, surely there is none in which its action is more obtrusive than in the history of human science. (MS 837: 3)

In the manuscript, Peirce does not really answer the question whether the evolution of reasoning is rather Darwinian or Lamarckian. More importantly, he does not specify what he understands by this or that "method of evolution". So, it would be inappropriate even trying to answer this question for Peirce before clarifying what the understandings of Darwinian and Lamarckian evolution actually were for Peirce. This task is addressed in the following chapter. For the moment I leave the question unanswered.

Without specifying the mechanism of evolution, we can conclude that in either case, what Peirce notices is that the nature of the "evolution" of logic is not limited to a quantitative development, but rather something critical happened in the development of human thinking which led to a revolution in the very core of logical thinking. Growth, the law of nature, evolution and finally *semeiosis*, is a dynamic process of never-ending differentiation and becoming, the taking of new forms. This process takes place thanks to direct and indirect interactions with the environment, creating interpretive chains of unlimited and potentially infinite representations of the already represented. Evolution and growth are two basic bridging terms between biology and the human sciences in Peirce.

Growth Extended: From Biology to Reasoning

Now let me develop on the topic treated in two already mentioned manuscripts of Peirce (MS 837 and MS 870) and in his paper *Basis of Pragmaticism*. In these three texts the main unifying idea can be summarized as growth representing what defines semeiosis and this is present in biological and cultural phenomena as well. Reasoning can be understood as a boundary of actuality between the biological and the cultural. It also depends on the school of thought whether we attribute reasoning more to nature or culture. For Peirce, this peculiarity is positioned elsewhere, closer to the questions of *What is reasoning* and *What is mind*. If the reader is familiar with Peirce's concept of quasi-mind (definition of quasi-mind e.g. in the text *Prolegomena to an Apology for Pragmaticism*, see Peirce 1906) omnipresent in nature, the question whether reasoning is natural or cultural somehow becomes superfluous. Many scholars agree upon the fact that reasoning is related to thirdness and symbolic representation, yet this assumption might lead to rather different conclusions. For instance, in the work of Terrence Deacon (1997) reasoning is derivative from symbolic representations, as defined by a hierarchical scaffolding from previous forms of indexical representations, which are themselves scaffolds from iconic representations. In Peirce we don't find any such gradual derivations of types of signs from other types of signs. Of course, icons are related to firstness, indexes are related to secondness and symbols are related to firstness in terms of qualia. We can talk about a hierarchy of qualia of firstness, secondness and thirdness, but there is no hierarchy between the aforementioned semeioses types since they represent only one of the aspects of Peirce's famous sign classification, specifically the relation between representamen and object. For Peirce, mind is also related to thirdness but not exclusively to symbolic representations. Peirce defines mind as thirdness in terms of habit making, self-control and growth in Law of Mind: "all the regularities of nature and of mind are regarded as products of growth" (Peirce in Bisanz 2009: 82). Moreover, if we allow hierarchical thinking in Peirce, it would go rather in the opposite directions, that is, from thirdness to secondness and firstness. This will be demonstrated in Chapter 4 with the NonReduction Theorem and impossibility of construction of a triad using only dyads or monads. Thirdness is when two different secondnesses are brought together, yet there is no claim of the temporal or hierarchical precedence of the two secondnesses preceding the thirdness. This is just another formulation of the statement that signs are but interpretations of other preceding signs. Semeioses are representations of an idea-potentiality, and we cannot access idea-potentiality in any other way than through semeioses. Yet, if the idea-potentiality remains infinitely in the chain of semeioses and does not transform into anything else but symbols and more symbols, for Peirce, this is not semeiosis in its full sense. Idea-potentiality should become, through the action of semeiosis (and translating from one symbol into another) an action, or a habit. Only when a habit is created can semeiosis be fulfilled. The same mechanism applies to growth in biology. A seed (a hereditary potential) is a potentiality and it can be translated from one generation to another by simple copying of the genetic information, but this is not a full semeiosis; therefore, we need epigenetics and evolution (evolutionary developmental biology), to transform the hereditary potential into an action of habit, an evolutionary habit, which is not a mere transferring of the same information from one generation to another. In the same way, the translation of a word from one language into another (interpreting a symbol into another symbol) is not yet a semeiosis. Semeiosis should result in action. By analogy, we can say that ideas grow as organisms grow by the force of the habit of action:

> But why should this idea-potentiality be so poured from one vessel into another unceasingly? It is a mere exercise of the World-spirit's *Spiel-trieb*, – mere amusement? Ideas do, no doubt, grow in this process. It is a part, perhaps we may say the chief part, of the process of the Creation of the World. If it has no ulterior aim at all, it might be likened to the performance of a symphony. The pragmaticist insists that this is not all, and offers to back his assertion by proof. He grants that the continual increase of the embodiment of the idea-potentiality is the *summum bonum*. But he undertakes to prove by the minute examination of logic that signs which should be merely parts of an endless viaduct for the transmission of idea-potentiality, without any conveyance of it into anything but symbols, namely, into action or habit of action, would not be signs at all, since they would not, little or much, fulfil the function of signs, and further, that without embodiment in something else than symbols, the principles of logic show there never could be the least growth in idea-potentiality. (Peirce in *The Basis of Pragmaticism*, in Bisanz 2009, 277)

In biology organisms create habits in order to evolve, develop or pass new acquired knowledge to future generations. "Symbols grow. They come into being by development out of other signs, particularly from icons, or from mixed signs partaking of the nature of icons and symbols." (CP 2.302) Even though in the quotation from "A Guess at the Riddle" cited above Peirce talks about mental signs, the principle of growing symbols equally applies to biology and biological signs. The most typical example of symbols in biology is surely the genetic script. Due to its arbitrary and conventionalized nature, the genetic script is but a result of a long evolutionary code establishing growth from previous non-symbolic forms of communication at the beginning of life on Earth, still growing and encoding new forms and new structures. The script itself grows: in other words, it is not immutable, because of the possibility of epigenetic marks or various possible forms of its "reading" or interpretations.

The Role of the Triad in Biology

This short excursion into Peirce's evolutionary theory of growth has demonstrated its indispensable triadic character. Evolutionary theory is necessarily triadic in the complex of Peirce's cosmology: chance and law are completed by habit. These principles of evolutionary development (evo-devo) are to be found also in other parts of his biological inquiry. To the triadic understanding of physiology and biological development Peirce dedicated his essay *A Guess at the Riddle* (published as MS 909, and in EP1) where, in addition to the biological application, the triad is explained as a basic principle of psychology and physics.

The following chapters of this book are dedicated to the triadic understanding of biological macromolecules, mostly proteins and protein structures. In this book, I will show how to apply the triadic relational understanding of Peirce's semeiotic to the biological macromolecules and phenomena in which they are involved. With the hope that triadic logic might illuminate some already known facts about protein structures, I will illustrate my argumentation with the aid of diagrams (Betagraphic). So to speak, my aim is to anchor triadic logic in contemporary biology. To close this chapter, let me restate the main principles of triadic logic and semeiosis as it was proposed by Peirce.

Peirce was interested in physiology mostly in two directions: firstly, in the processes of the nervous system and the activity of nerve cells; secondly, in the protoplasm[7]. The threefold nature of the nervous system cell is, according to Peirce, based upon the following three principles: "first, the excitation of cells; second, the transfer of excitation over fibres; third, the fixing of definite tendencies under the influence of habit" (EP1: 266).

The taking of habits of a nerve cell is more concretely described in the section What is Love? of this book. As stated in A Guess at the Riddle, protoplasm behaves similarly to a nerve cell, that is, protoplasm is equally based upon the three precepts, of chance as firstness (a random excitation as sensation of a pure feeling), law of the transfer as the secondness (relation between the pure feeling and something to which this feeling might relate), and habit as thirdness (fixing the tendencies for the future cases of excitations). In the case of protoplasm, these correspond concretely to sensibility (first), motion (second), and growth (third).

It is clear enough that for Peirce triadic relations are analogous structures across various disciplines, or likewise in various biological processes, whether

---

7   As Santaella Braga points, the protoplasm was only correctly defined after Peirce's death
     (Santaella Braga 1999: 5–21).

they be connected to a single individual organism, a cell or an organ, or to the whole evolutionary process. To put in in other words, we can quote from another Peirce's essay:

> In biology, the idea of arbitrary sporting is First, heredity is Second, the process whereby the accidental characters become fixed is Third. Mind is First, Matter is Second, Evolution is Third. (Peirce, *The Architecture of Theories*, in Bisanz 2009, 68)

Here we have a somewhat different explanation of the triad in biology where Mind and Matter are taken into relation with habit (evolution). In the sentence cited above the whole semeiotic of Peirce is explained by a beautiful simplicity of only a few words. Mind is firstness, matter is secondness, and what actually connects mind and matter is the evolutionary habit as thirdness, something necessarily situated in space and time (because evolution is only thinkable in terms of time), and something equally applicable to both evolution and physiology.

Chapter 2 Triadic Evolution

Method of Interdisciplinary Biology

Charles Sanders Peirce was first of all a scientist who greatly contributed to the development of physics, mathematics, logic and chemistry. Philosophy was for him only one of a number of sciences, and in his classification of sciences it occupied the second position. Thus, the general association of Peirce with philosophy is not precisely correct. Of course, his contributions to philosophy are undeniable, yet we must understand Peirce's philosophy as only one part of his scientific contribution and we should specify it as a natural philosophy the aim of which is to understand nature and the cosmos. With this optic, Peirce's writings on biological phenomena only complete the overall ensemble of Peirce's opera. Understanding Peirce's hypotheses on biology helps us to comprehend his cosmology and the whole of his scientific enterprise. The figure of C. S. Peirce as contributor to the scientific and philosophical discussion of evolutionary theories of the 19th century will be the central part of this chapter. Peirce commented on evolution in many parts of his writings[8] even though none of these papers were fully dedicated to evolution: evolution was in many respects for Peirce only a complementary part of a larger cosmology. I say "only", but the non-exclusive focus on evolution in Peirce shouldn't be viewed as a devaluation of its vigor. It is in the interconnectivity of Peirce's works from different fields that his originality and magnificence dwells.

Semeiotic is a relational logic that also defines the scientific method of Peirce, finding relations among different disciplines and fields, creating analogies and formulating hypotheses. Peirce invented his logic of relatives by analogy from chemical radicals, which itself is an analogy for verbal valency. This is only one instance from a plethora of examples demonstrating how Peirce was a veritable interdisciplinary scientist (Ketner 2009). With this complex image we can approach his evolutionary theory. My aim behind this brief excursion into Peirce's theory of evolution is to introduce Peirce's attitude towards biology as a discipline in general, contextualizing some of the sub-disciplines and contemporary achievements of biology from a larger perspective. This might be a difficult task, given the time gap between the state of biology as a scientific discipline in

---

8   Most of these were between 1891 and 1893, published in *The Monist*. For the complete papers see Bisanz (2009). For the evolutionary theories see also Burks (1997).

Peirce's era and today's situation in biology, with all its enormous scales of sub-disciplines and interdisciplinary research, such as biochemistry, biophysics, bio-informatics, genetics, biological engineering, etc.; my hope is to demonstrate the universality of Peirce's ideas not only horizontally in the interdisciplinary sense, but also vertically in the chronological sense.

In the lifetime of Peirce, evolutionary theory was in the center of scientific discussion (Darwin's *Origin of Species* appeared in 1859). Therefore, it is not surprising that Peirce was tempted study this subject. He was also interested in the behavior of protoplasm (also called in some of his writings "albuminoids") and neurology in the proprieties of the cells of the nervous system. Among the three famous ways of reasoning, induction, deduction and hypothesis or abduction, also the natural sciences are classified according to a particular way of reasoning applied in a given scientific field. In this classification, Peirce believed that biology was a science of abduction:

> Of the natural sciences, we have, first, the classificatory sciences, which are purely inductive – systematic botany and zoology, mineralogy and chemistry. Then, we have the sciences of theory, as […] astronomy, pure physics, etc. Then, we have sciences of hypothesis – geology, biology, etc. (EP1: 199)

Understanding biology as an abductive science makes it even more tempting to try to illustrate the intersection between contemporary biology and Peircean semeiotic theory. Abduction being the most relevant scientific method of reasoning and being the most suitable method for evolutionary theory, it rests upon at the same time the pattern for semeiosis. With the hope that Peirce's understanding of evolution might serve as a preliminary part for later applications to epigenetics and protein folding, I will try to illustrate the ways in which Peirce's evolutionary theory differ from the more famous evolutionary theories of his predecessors or contemporaries. While we have some direct comments by Peirce on Darwin and Lamarck, a comparison with Henri Bergson and his creative evolution might bring some new insights to the theory of Peirce's *evolutionary love*. After merging biology and Peirce's abductive method, I will conclude this chapter by introducing the notion of triadic relational logic in biology.

To begin, it is necessary to localize the place of biology in Peirce's classification of sciences, and to study Peirce's definition of biology in detail. I will adhere to the definition by Peirce from the *Century Dictionary* (Whitney 1889–1891) in the case of the definition of biology but also in other parts of this chapter and book. CD provides definitions formulated by Peirce (he was author of about 6000 entries and editor of many others for the CD). It is a useful tool in understanding not only Peirce's ideas, but also the way he decided to present them to the general and scientific public. To understand Peirce's definition of biology, we will not go, surprisingly, to the entry "biology". More interesting are his notes on biology under the entry "science", especially for the aforementioned inseparability of biology from the rest of the scientific disciplines.

> In reference to their degree of specialization, the sciences may be arranged as follows. (A) Mathematics, the study of hypothetical constructions, involving no observation of facts, but only of the creations of our own minds, having two branches – (1) pure mathematics [...] (2) applied mathematics (B) Philosophy, the examination and logical analysis of the general body of facts – a science which both in reason and in history precedes successful dealing with special elements of the universe – branching into (1) logic and (2) metaphysics. (C) Nomology, the science of the most general laws or uniformities, having two main branches – (1) psychology and (2) general physics. (D) Chemistry [...] (E) Biology, the study of a peculiar class of substances, the protoplasms, and of the kinds of organisms into which they grow. (F) Sciences of organizations of organisms, embracing (1) physiology [...] (2) sociology [...] (G) Descriptions and explanations of individual objects or collections, divided into (1) cosmology [...] (2) accounts of human matters. (CD: 5397, accessed 25/2/2024)

The first thing to notice is that in his classification of sciences Peirce attributes the first position to mathematics, after that the science of philosophy follows, then biology comes only after physics and chemistry. To explain this order of the sciences it is important to understand what the general meaning of the word "science" was in the times of Peirce. In this connection, Markoš and Švorcová in their recent book on epigenetics write:

> Science, as many scientists stress and maintain up to our times, longs after the existence of a Platonic element in nature, the basic, atemporal, level of description. One of the founders of molecular biology, J. Monod, states: "In science there is and will remain a Platonic element which could not be taken away without ruining it. Among the infinite diversity of singular phenomena science can only look for invariants." (Monod 1971) [...] Darwin's essential contribution thus lies in **not** being faithful to any such Platonic ideal: history definitely is not Platonic. In his treatise on the history of biology, Emanuel Radl (1909, English translation 1930) perceived the problem clearly: "Since about 1890 practically no advance has been made in our knowledge of the origin of living organisms. Can this, by any chance, be due to the fact that our methods have been too exclusively the methods of science? It would be quite possible to treat the history of the world of living organisms as an historical problem, and to attempt to solve the problem by historical methods. It would then be analogous to the study of the origin of religions or of the various European states. The more I, personally, think on the subject, the more I believe in this method of attack." (Radl 1930) (Markoš and Švorcová 2019: 63)

What probably remains in scientific reasoning up to the present day, as the authors note, is a search for Platonic forms. Probably no contemporary scientist would agree with this claim, however extreme the allusion to Plato might seem, science nevertheless does seek immutable universalities, laws, ideals, and invariants. Biology is situated in Peirce's classification closer to the humanities rather than to mathematics and physics. This is the point where Peirce's understanding of biology is close to that of Markoš. Biology does not seek (not in the same sense as in physics, at least) universal immutable laws, or if it does seek these laws they have still not been discovered, except perhaps those of the genetic code. Human cultural phenomena are dynamic and always changing in the course of history. Biological phenomena are close to the humanities in this sense: it is impossible to define them in their complexity with a closed set of immutable rules/laws. What brings biology closer to the humanities on the classificatory scale is its methods. The crucial part of the quote is the mention and citation by Emanuel Rádl, Peirce's contemporary Czech biologist (1873–1942) who hypothesized a fruitful and happy application of the method of history to the study of organisms. Historical or archeological inquiry uses methods of tracing the past through partial messages we can find left by material objects or the tracks that remain after objects. The exact same method is used in evolutionary theory, with the difference that the inquired object is not cultural but natural phenomenon.

> We may study an animal as thoroughly as we please; we may compare it in the most detailed manner with other forms; but this will not enable us fully to understand its nature for it bears within it the traces of the past, traces which can only be revealed by historical study. (Radl 1930: 174, cited from Markoš and Švorcová 2019: 64)

Markoš and Švorcová's Monod-inspired criticism of the scientific method shouldn't be understood as a disavowal of the scientific approach. What they want to stress is the importance of the difference between biology on the one hand, and the natural sciences based exclusively on physics or chemistry on the other hand. Biology deals with living organisms undergoing development and changes both at the individual (but even micro-individual-genetic) and the phylogenetic level. Peirce proposed a solution to this radical positioning of biology opposite to the scientific method. Thanks to Peirce's definition of scientific method we are able to merge evolutionary theory and the phenomenon of life with the scientific approach. According to Peirce, scientific method can also be abductive, hypothetical. And abduction is exactly the biological method. Scientific method can lead towards the definition of immutable laws but can also lead towards the definition of dynamic and evolving natural habits. What Markoš and Švorcová criticize about the method in biology, and what they find Darwin to have successfully achieved in this theory is a historical awareness for understanding biological processes as narratives.

This kind of methodological fusion between the humanities and natural sciences recalls the centuries-old Humboldtian ideal of education recently developed by Arran Gare. According to Gare, a revolution in scientific thought is needed

> that requires us to alter to some extent our understanding of what science is, and that this should have been undertaken long ago. This revolution was called for by Friedrich Schelling at the end of the Eighteenth Century in his effort to develop a philosophy that would unite science, history and the arts. The failure to fully carry through this revolution, required to make life and mind intelligible, is now hindering not only the advance of biology (and medicine), but all science, and also, the humanities. It is also damaging major economic, social, political and cultural institutions, including universities which have abandoned the Humboldtian model and ideal of education that inspired academics in the sciences and humanities for nearly two hundred years, and blocking effective action in response to the global ecological crisis. (Gare 2019: 34)

The fusion between science, the arts and history is not to be understood as unscientific, but on the contrary. To fulfill this ideal, a reconsideration of the scope of science is needed. If we consider today's general understanding of science and the scientific method, the application of any historical approach to biology can only delegitimize its position. Yet recent developments in epigenetics, evo-devo and similar biological branches calls for a historical, processual, archeological[9] and dynamical approach. Life has become, thanks to experimental evidence from the recent decades, no longer exhaustively describable with purely immutable laws. In the context of epigenetics, life is compared to a chess game "without beginning and end" (Markoš and Švorcová 2019: 188), a game that goes on in a constant dynamic development.

Chance, Law and Habit as Reciprocal Components of Evolution

The evolutionary theory of Peirce meets his scientific and mathematical understanding of the universe in a vast array of relations between mind and matter, potentiality and actuality, individuality and continuity. Thus, the evolutionary theory of Pierce proceeds in direct continuity with the concepts of chance, law, and habit also applied to other branches of science or other scientific disciplines approached by Peirce. Chance, law and habit are three basic components of Peirce's evolutionary theory, they are defined by mutual reciprocity, as will be explained further and are by their nature interdisciplinary, or universal, that is, they are not limited purely to evolution: the notion of chance stems from

---

9   See Ariza-Mateos et al. (2019).

mathematics, the notion of law emerges from physics and the notion of habit is related to cosmology. As a consequence, the evolutionary theory by Peirce is naturally related to the whole of his scientific writings, which include his science of philosophy. It would be improper to consider the evolutionary theory of Peirce separately from the rest of his achievements. Only in connection with logic and semeiotic is it possible to comprehend Peirce's evolutionary theory and vice versa, the evolutionary theory completes the whole scientific work of Peirce. The three terms of chance, law, and habit, central for evolutionary theory, are found also in Peirce's other doctrines. Applied to evolutionary theory they acquire a new form and a new nomenclature, becoming accordingly tychasm, anancasm, and agapasm. In the following section of this chapter, I will focus accordingly on the three aforementioned notions, one after another. Tychasm, anancasm, and agapasm are the "three modes of evolution" (Peirce in Bisanz 2009, 126). Even if these are called "modes of evolution", it could be said that they are but singular components of a unique evolution, of the evolutionary process of dynamical development consisting in the fixing of acquired traits and creating of new ones. Chance, law, and habit are necessary components of evolution, which is only possible when the three components complement each other.

Chance:

The first type of evolution, evolution by Chance, is a response to Darwin's principle of random variation as the driving process of evolution of species, in Peirce's terms described as "fortuitous variation" (Peirce, *utionary Love*, in Bisanz 2009, 122). The insertion of the term chance into evolutionary theories was Darwin's revolutionary contribution to science. It was in direct contradiction with the preceding law-like search for Platonic forms in organic morphology. In the CD, definition n. 8 of the entry "Chance" says:

> Probability, the proportion of events favorable to a hypothesis out of all those which may occur: as, the chances are against your succeeding.

Definition n. 9 continues:

> Fortuity, especially, the absence of a cause necessitating an event, or the absence of any known reason why an event should turn out one way rather than another, spoken of as if it were a real agency, the variability of an event under given general conditions, viewed as a real agency. [...] **Absolute chance**, the (supposed) spontaneous occurrence of events undetermined by any general law or by any free volition. According to Aristotle, events may come about in three ways: first, by necessity or an external compulsion, second, by nature, or the development of an inward germinal tendency, and third, by chance, without any determining cause or principle whatever, by lawless, sporadic originality. – By

chance, without design, accidentally[…] **Even chance**, probability equally balances for and against event. – **Main chance**, the chance or probability of most importance or greatest advantage: hence, the end or stake to be kept most in view, the chief personal advantage […] Theory or doctrine of chances. See probability.

"Chance" is defined in opposition to law in nearly all of the sub-meanings and at the end of this quote. The fact that chance is opposed to law does not however mean that it cannot be related to a certain type of rules or be mathematically described. Actually, the term "chance" originates from mathematics and Peirce treated mathematical Chance in many papers in connection with statistics. We can see the mathematical definition of chance in the sense of probability. Peirce indicates the entry for Probability (CD: 4741, accessed 25/2/2024), which says, in definition n. 2 of the entry:

> Quantitatively, that character of an argument or proposition of doubtful truth which consist in the frequency with which like propositions or arguments are found true in the course of experience. Thus, if a die be thrown, the probability that it will turn up ace is the frequency with which an ace would be turned up in an indefinitely long succession of throws. It is conceivable that there should be no definite probability: thus, the proportion of aces might so fluctuate that their frequency in the long run would be represented by a diverging series. Yet even so, there would be no approximate probabilities for short periods of time. All the essential features of probability are exhibited in the case of putting into a bag some black beans and some white ones, then shaking them well, and finally drawing out one or several at random. (CD: 4741, accessed 25/2/2024)

For Peirce, the study of chances is part of the mathematical doctrine of probabilities and represents one of the most important components of his early writings (for instance, papers from 1878 published in *Popular Science Monthly*: *The Doctrine of Chances*, *The Probability of Induction*, *The Order of Nature*, but also papers published in *The Monist* and republished as 2009 Bisanz (ed.)).

Peirce also commented on the principle of chance in physics, on the idea of primordial chaos:

> In fact, chance is but the outward aspect of that which within itself is feeling. I long ago showed that the real existence, or thingness, consists in regularities. So, that primeval chaos in which there was no regularity was mere nothing, from a physical aspect. Yet it was not a blank zero, for there was an intensity of consciousness there in comparison with which all that we ever feel is but as the struggling of a molecule or two to throw off a little of the force of law to an endless and innumerable diversity of chance utterly unlimited. (Peirce, *Man's Glassy Essence*, in Bisanz 2009, 114)

In line with modern physics and quantum physics theory, Peirce agrees to see chaos and pure chance as the primary state of the universe. Evolution by Chance, or tychasm, explains how the principle of random changes might be part of the evolution of species. But, as Brier points out (Brier 2008a), Peirce's idea of chaos is not deterministic. Chaos cannot be considered an absence of law, if the law arises from chaos. It is not a "blank zero". If chaos is primary in the universe, we need to explain law from chaos and not vice versa. As pointed out by Scott, chance is not an absence of law, but is governed by rules, these rules being expressible by statistical law. Chance determines change and self-control, which is very important for Peirce's notion of the phaneron. Chance together with continuity and the absolute are three mathematical notions determining the universe, that is to say, arts and nature without a difference (Scott 2006: 51).

To explain law from chaos is to find out how laws may emerge from something which is but "an intensity of consciousness". This statement might seem to a certain degree paradoxical. Most of our understanding of consciousness depends on order, and from a materialist viewpoint it would be a highly ordered matter, a matter of neurons ordered in a certain way giving birth to consciousness. What is so different in Peirce in comparison with this rather classical understanding of consciousness is that for him, consciousness is not the final result of the ordering of matter, but the original starting point. Therefore, Peirce disagrees in principle with the materialistic understanding of consciousness (Peirce, *The Architecture of Theories*, in Bisanz 2009, 58–69). Consciousness was at the beginning of everything. Peirce breaks down the Cartesian dualism, but in a very specific way which is inseparable from his pantheistic or pansemiotic view on the world. The pansemiotic viewpoint calls for primordial consciousness or mind which precedes its material embodiment[10], or better, the two of them, mind and matter, evolve in parallel (Burks 1997). We can understand this primordial consciousness as a firstness, as a *feeling of an idea*. In the words of Søren Brier, "Peirce viewed mind and matter as connected in a continuum, and that matter has some internal living qualities" (Brier 2019: 69).

In this way chance is understood in the context of Peirce's evolutionary theory. Chance is not chaos, chance is spontaneity from which laws are created. Peirce accounts for a "chance in the form of a spontaneity which is to some degree regular [...] a spontaneity which develops itself in a certain and not in a chance way" (CP 6.63).

---

10   As Burks comments (1997: 531–532), Peirce's evolutionary theory is often mistakenly interpreted as panpsychic, which is incompatible with today's scientific paradigm. Burks says that Peirce's theory is the best non-reductive philosophical theory of evolution. Obviously in Burks non-reductive and panpsychic are two categories which are disconnected and so we should also understand Peirce's evolutionary theory.

In his essay *The Logic of the Universe* Peirce even mentions the initial potentiality, the plurality of options that leads to the creation of habits.

> The evolution of forms begins or, at any rate, has for an early stage of it, a vague potentiality; and that either is or is not followed by a continuum of forms having a multitude of dimensions too great for the individual dimensions to be distinct. It must be by a contraction of the vagueness of that potentiality of everything in general, but of nothing in particular, that the world of forms comes about. (CP 6.196)

Law:

As a complement to evolution by Chance, evolution by Law is proposed next by Peirce. In the CD entry for "Law" definition n. 3 reads as follow:

> a proposition which expresses the constant or regular order of certain phenomena, or the constant mode of action of a force, a general formula or rule to which all things, or all things or phenomena within the limits of a certain class or group, conform, precisely and without exception. (CD 3375, accessed 25/2/2024)

"Without exception" is a crucial term in the definition of Law. Evolution by law is understood within the context of the philosophy of necessitarianism. "Mechanical necessity" (Peirce, *Evolutionary Love*, in Bisanz 2009, 124) more effectively explains what Peirce expresses by the principle of Anancasm. Peirce proposed the word Anancasm to nominate the principle of necessity or law in the context of the evolutionary mechanisms of his three modes of evolution. "Law" here is to be comprehended in terms of physical laws, so that any step in evolution would be explained by a physical, chemical or environmental necessity. Necessity is in part driven by causal process, causality, in contrast to the randomness of the first type of evolution. It develops into the central mechanism behind the second type of evolution. Causality in anancastic evolution is development by the mere force of circumstances or of logic. Causality is dependent exclusively on "the pressure of external circumstances or cataclasmine evolution" (Peirce, *Evolutionary Love*, in Bisanz 2009, 131). Even more radically we could say that Anancasm as causality is in contradiction with the teleological understanding of evolution. Probably also tychastic evolution should be defined in terms of being opposed to a goal-directed development. Lamarck is mentioned in connection to Anancasm as well.

To illustrate the anancastic principle in Lamarck, we can refer to the chapter *Influence of Environment* from his *Zoological Philosophy* (Lamarck 1963: 107–127). See the following illustration:

> Nevertheless, some of these herbivorous animals, especially the ruminants, are incessantly exposed to the attacks of carnivorous animals in the desert countries that they

inhabit, and they can only find safety in headlong flight. *Necessity* has in these cases *forced them* to exert themselves in swift running, and from this habit their body has become more slender and their legs much finer; instances are furnished by the antelopes, gazelles, etc. [...] It is interesting to observe the result of habit in the peculiar shape and size of the giraffe (Camelo-pardalis): this animal, the largest of the mammals, is known to live in the interior of Africa in places where the soil is nearly always arid and barren, so that it is *obliged to* browse on the leaves of trees and to make constant efforts to reach them. From this habit long maintained in all its race, it has resulted that the animal's fore legs have become longer than its hind legs, and that its neck is lengthened to such a degree that the giraffe, without standing up on its hind legs, attains a height of six meters. (Lamarck 1963: 122)[11]

The lexicon of mechanical necessity might be found in lexemes such as "necessity", and "obliged" in the above quotation. Interestingly enough, Lamarck's doctrine is paradoxically often described as teleological or vitalistic, or even creationist, due to the fact that causality is in many explanations of Lamark's theory directed towards a specific scope, for example the famous prolongation of giraffe's neck. Yet there are many supporters of Lamarkism who are actually materialistically oriented, as Švorcová points out (Švorcová 2012: 22). Moreover, as Švorcová states elsewhere, "Lamarck did not introduce a specific, elaborated theory of the inheritance of acquired characters. He explained the variability of organisms on a behavioral basis as part of the law of use and disuse" (Markoš and Švorcová 2019: 64). Thus, Lamarck's theory is rather causalistic and behavioral than vitalistic, contrary to the telological explanations of Lamark's doctrine. As Markoš and Švorcová propose, we can talk about teleology in Lamarckian theory, yet this teleology is a mechanistic or necessitarian teleology, a lawful form of teleology closer to mechanistic than teleological explanations.

We get from Lamarck a concept of crystallization of a kind, which takes place whenever conditions will allow it [...] Crystallization is a physical, lawful process leading **always** to the same set of the appearances of life as its outcome [...] The physical basis of Lamarckian teleology comes to the fore when he speaks about the sequences of events: all forms of, say animals, strive towards, or better, are predestined to give rise-after generations-to human beings. In short, for Lamarck, if all mammals would go extinct, they will "grow" again from lower forms of life, reconstituting previous forms including humans. (Markoš and Švorcová 2019: 62)

What might be misunderstood here is the very mechanism of causality as formulated by Lamarck. According to Lamarck (1963: 122), it is not to be understood in a teleological manner; rather, there is a direct causality going in the direction

---

11   Italics inserted by the author of the book.

cause-effect, where cause is the experience of parents and effect is a prolongation of the neck of the offspring. Thus, the only driving force of evolution is the mere force of circumstances as a blind play of logical necessity. See more commentaries on Lamarck in the penultimate section of this chapter. To an even better understanding of the notion of law as present in Lamarck, let me conclude with a quotation by him expressing explicitly the idea that life is nothing but mechanical necessity depending on physical laws:

> It seems to me that it was much easier to determine the course of the stars observed in space, and to ascertain the distance, magnitudes, masses and movements of the planets belonging to our solar system, than to solve the problem of the origin of it, life in the bodies possessing and, consequently, of the origin and production of the various existing living bodies. However difficult may be this great enquiry, the difficulties are not insuperable, for in all this we have to deal only with purely physical phenomena. (Lamarck 1963: 184)

Habit:

For Peirce, the two already existing views on evolution, correspondingly represented by the theories of Darwin and Lamarck were completed by the third one: evolution by habit.

Habit is defined in the CD entry by direct connection with biology and even with the notion of growth, introduced in the first chapter of this book. Mostly the first and the second meaning of the entry are relevant for Peirce's evolutionary theory.

First meaning:

> A usual or characteristic state or condition, natural condition, attitude, appearance or development, customary mode of being. Specifically [...] (b) in zool. and bot. the general aspect and mode of growth of an animal or plant.

Second meaning:

> A usual or customary mode of action, particularly, a mode of action so established by use as to be entirely natural, involuntary, instinctive, unconscious, uncontrollable, etc. : used especially the action, whether physical, mental, or moral, of living beings. (CD: 2673, accessed 25/4/2024)

As Peirce states in *Doctrine of Necessity Examined*, it is equally naive to believe that evolution is driven exclusively by mechanistic laws as to believe that it is driven by a pure chance. The word "exclusively" is noteworthy here. It implies that mechanistic laws and chance are actually part of evolution, but are not the only ones in the evolutionary process. There is an interplay between Chance, mechanistic laws and Habit:

It would suppose that in the beginning, infinitely remote, there was a chaos of unpersonalized feeling, which being without connection or regularity would properly be without existence. This feeling, sporting here and there in pure arbitrariness, would have started the germ of a generalizing tendency. Its other sportings would be evanescent, but this would have a growing virtue. Thus, the tendency to habit would be started; and from this with the other principles of evolution all the regularities of the universe would be evolved. At any time, however, an element of pure chance survives and will remain until the world becomes an absolutely perfect, rational, and symmetrical system, in which mind is at last crystallized in the infinitely distant future. (Peirce, The Architecture of Theories, in Bisanz 2009, 69)

In this quotation chance as "feeling" is explained, but also it is specified how the notion of chance relates to the notion of Habit. As a matter of fact, the notion of habit represents the basics of the whole cosmology of Peirce. He uses both terms chance and law as the driving forces of evolution but only when habit enters the evolutionary process, evolution might be completed. Tychasm and Anancasm are completed by Agapasm. Agapasm is a term deriving from Greek lexeme for "love". It might seem that Peirce had some romantic tendencies when introducing this term to his evolutionary theory, yet the meaning of the word is much more complex than that.

What is Love?

Evolutionary Love is an article which appeared in *The Monist* and was reprinted in the book *Chance, Love, and Logic* (see Fig. 1). In this essay Peirce defines *agápē* as ἀγαπη and ἀγαθον, in Greek "love" and "loveliness", which means the force of an attraction between people, ideas or things. Surely it is difficult to understand whether the use of the term love was intended as metaphorical and, if yes, in what sense. Peirce maintains that the English translation is not adequate, as he says, this "pirate-lingo" (Peirce, *Evolutionary Love*, in Bisanz 2009, 117) has no adequate words to express the meaning of *agápē*, therefore probably the use of the word is not intended to be metaphorical, but merely was used due to the lack of opportune terminology.

Simplified, *agápē* is the habit to love, which the universe has the tendency to take. But what does it really mean to say that the universe has the habit to love? For Peirce there was also a religious argument taken from the Christian faith, but he does not really stress the theological explanation of evolution. For Peirce, the nature of the universe to take habits derives from the specific understanding of love as an overwhelming principle in the universe where there are no contradictions to love: hate is only a lack of love, so the physical forces of attraction in the universe are only this kind of *love*, and attractions are kinds of relations of any sort between things or ideas. Agapasm has many forms, from physical attractions manifesting in the form of physical laws to various modes of affection in human culture. In physical forms, such as the force of gravity for instance, the force of attraction might be qualified as genuine anancasm or degenerate agapasm, because of its lawfulness typical for anancasm. The frontier from anancasm to agapasm is not represented by a strict line; some critical phenomena might perfectly be attributed to both anancasm and agapasm. The tendency to take habits might result, in extreme cases, in a final interpretant which is hardly distinguishable from law, this being the case of the frozen rules of the genetic code, created by habit making but resulting in a law-like frozen rule. It is also very useful to apply the distinction between soft and hard habits by ISP 2019 (PS 10):

> We shall understand *Habit* as an activity relation of a particular sort between two components or phenomena—here designated as *items*—such that the relation, described by the proposition "If *item one*, then *item two*," holds to some degree, as shown by experiment. The notion of *degree* can range over varying strengths of relationships, such as "a small degree of likelihood," to "will be the case regularly and reliably." We designate habits of the latter kind as hard habits, while those of various weaker degrees we describe as soft habits. Efficient causal relations, or "laws of nature" are examples of soft habits. (ISP 2019: 28–29)

Whereas "laws of nature" are hard habits, activity relations in biology can be both hard and soft habits (as demonstrated by ISP, see also Ketner 2023: 14). In some cases, the distinction between hard and soft habits might not be easy. The major problem of the competing frameworks of the Modern Synthesis on one hand and EES on the other hand lies indeed in describing the phylogeny and ontogeny in terms of hard habits exclusively (modern synthesis) or in combination of both hard and soft habits (EES). Hard habit is the result of actions of soft habit, like anancasm is the result of the action of agapasm:

> Herein, it must be admitted, anancasm shows itself to be in a broad acception a species of agapasm. Some forms of it might easily be mistaken for the genuine agapasm. (Peirce, Evolutionary Love, in Bisanz 2009, 127)

It is important to note that the frontier between anancasm and agapasm corresponds to the logical relations of Peirce in the way that the direction goes from agapasm to anancasm and not the other way around. Agapasm corresponds to thirdness as a category representing mediation; as the mediation between secondess and firstness, it has a relational character, but the relation is a triadic one which cannot be reduced to dyadic relations of anancasm. Thus, laws and randomness in nature are derivatives of agapasm, from evolutionary love.

Love is thirdness in the sense of the third mode of evolution which proceeds by habit making, and love in this meaning is but a synonym for Growth (see Chapter 1). To illustrate the tendency of habit making in biology, we can take the example by Peirce of five principles of habit making in nerve cells:

> First, when a stimulus or irritation is continued for some time, the excitation spreads from the cells directly affected to those that are associated with it, and from those to others, and so on [...] Second, after a time fatigue begins to set in. Now besides the utter fatigue which consists in the cell's losing all excitability, and the nervous system refusing to react to the stimulus at all [...] Third, when, from any cause the stimulus to a nerve-cell is removed, the excitation quickly subsides. That it does not do so instantly is well known, and the phenomenon goes among physicists by the name of persistence of sensation [...] Fourth, if the same cell which was once excited, and which by some chance had happened to discharge itself along a certain path or paths, comes to get excited a second time, it is more likely to discharge itself the second time [...] This is the central principle of habit; and the striking contrast of its modality to that of any mechanical law is most significant. The laws of physics know nothing of tendencies or probabilities [...] Fifth, when a considerable time has elapsed without a nerve having reacted in any particular way, there comes in a principle of forgetfulness or negative habit rendering it the less likely to react in that way. (EP1: 264)

These five principles are illustrated by the example of a card game, how to obtain only spades by virtue of habit taking (CP 1.391). Now we can extend the five principles of the taking of habits by nerve cells to the evolutionary process, but also to other biological processes, at the level of molecular biology and the establishment of the genetic code and protein folding, which will be approached in the following chapters.

Some authors understand the agapasm as "creative love" (Acosta 2015), others as a law of mind to take habits (Brier 2006).

Peirce defines the evolution of the world in terms of diversity and uniformity, just as Stuart Kauffman (2000) talks about the subcritical state of the cells, which means the interface between constantly expanding biological diversity and freezing in the established habits. Kauffman's theory has been compared with H. Bergson's book *Creative Evolution* (Markoš 2003). I am of the opinion, however, that differently from Bergson, Kauffman shows points of contact with Peirce's philosophy of meaning formation by chance and habit, and with the definition of meaning as action in the world. Kauffman's description of the biosphere through the dissemination of diversity, incompatible with the second law of thermodynamics, which he presents in his book *Investigations* (2000), is almost identical to Peirce's ideas expressed in the *Doctrine of Necessity Examined*. Kauffman defines the biosphere as continuously expanding its future possibilities, while retaining a reasonable degree of "traditionalism" (habit) when organisms act in well-established trajectories. Too many and too rapid changes can become malignant for species; therefore, the harmony between chance and habit is crucial. In Kauffman, there is practically no difference between the notions of law and habit, and they are understood as one and the same principle. Kauffman's vicinity to Peirce resides also in that both scientists regard evolution and the notion of chance through mathematical theory on probabilities and statistics and concluded that besides chance and law there is another "law" which acts upon the future and is directed towards the expansion of diversity in nature. Kauffman's ideas are very close to what Peirce expressed in CP 6.58:

> there is agency in nature by which the complexity and diversity of things can be increased, and that consequently the rule of mechanical necessity meets in some way with interference.

Living organisms transmit by heredity the accumulated rules of survival and orientation in the world to following generations, providing them with the right way to interpret the world. For Peirce, such an important concept was consequently habit. Habit is a notion regarding semeiosis in general, not only the semeiosis taking place in the evolution of species or the cosmos. Habit is also present in semiotic theories on cognition as well. As C. Paolucci states, "for the sake of economics of labor, we tend to understand new meanings based on old unambiguous meanings" (Paolucci 2010: 382). Paolucci calls this process "interpretive comfort".

Teleology and Teleonomy

It has to be mentioned that some authors refer to another triad of evolutionary principles; that is, instead of tychasm, anancasm and agapasm they mention tychasm, *synechism* and agapasm (Burks 1997; Stjernfelt 2007) as the three main propulsive principles of evolution in Peirce. Let me examine this alternative in the context of evolution. Synechism is mentioned on p. 119 of *Evolutionary Love* (in Bisanz 2009) as a part of agapasm.

Some authors associate synechism with evolutionary theory by Peirce, yet synechism is not directly involved in his evolutionary theory in the sense that it does not take part in any important way in any of Peirce's writings directly dedicated to evolution, such as Evolutionary Love, The Architecture of Theories or others (the essays on evolution were published as the second part of the publication Chance, Love, and Logic entitled Love and Chance and include: The Architecture of Theories, The Doctrine of Necessity Examined, The Law of Mind, Man's Glassy Essence); synechism can be also understood as evolutionary growth as a natural tendency of the cosmos to a progressive and continuous growth directed to a given goal. This goal in the evolutionary understanding might be seen as a progressive perfection, trying to become better than before, therefore Peirce's evolutionary theory is sometimes referred to as teleological, in reference to both synechism and agapasm (Burks 1997, Brier 2008a). In this understanding, the teleological principle would be the third completing term to the principles of Chance and Law.

> Thus, in order to build a fruitful relationship between chance and law, teleology became necessary, and, in order to give a place to diversity and complexity, a conception of chance as a cause always operative in the world became necessary as well, this last form of chance is what Peirce also names "spontaneity." In fact, only from real spontaneity are evolution and change possible. (Acosta 2015: 37)

Teleology is a controversial term in scientific discourse, although in biology it is somehow difficult to completely omit. It must be clearly stated that there are several forms of teleology. I already mentioned the teleological mechanism in the case of Lamarckian evolution, where teleology might be and often is mechanistic, even if the two terms are often put in a contradiction (Markoš and Švorcová 2019), mostly in the sense of Lamarckian teleology. Yet Peirce's teleology is not a mechanistic one because it is related to the notion of habit. Peirce's teleology is specific and deserves more consideration.

Peirce relates the teleological principle to agapasm directly when describing the habit taking of a nerve cell:

> Thus we see how these principles not only lead to the establishment of habits, but to habits directed to definite ends, namely the removal of sources of irritation. Now it is precisely action according to final causes which distinguishes mental from mechanical action; and the general formula of all our desires may be taken as this: to remove a stimulus. Every man is busily working to bring to an end that state of things which now excites him to work. (CP 1.392)

As Burks (1997) remarks, in the last decades probably starting from Watson's and Crick's discoveries, there seems to be no space for an agapastic evolution in the sense of teleology or final causes. The genetic code is often understood as the result of blind chance concluding in the random pairing of amino acids with DNA bases. And yet for some scholars the genetic code is related to final causation:

> Organisms are governed by final causality in the sense of their tendency to take habits and to generate future interpretants of the present sign actions. Codes in living systems are correspondences based on final causation that cannot be inferred directly from natural laws. They are based on the formal causation of the protosemiotic differences and pattern fitting information mostly on the chemical level of interaction. The

physical interactions are based on laws and efficient causation of energy transfer. (Brier 2008a: 242)

But we do not need to perform time-travelling in order to demonstrate Peirce's uniqueness, which might be (wrongly) seen as anti-scientific. Apparently already at the time of Peirce teleology was not tenable as a scientific position. Peirce seems to feel the need to defend his teleological position in the scientific context. Let's have a look at his proper definition of teleology:

> It is [...] a widespread error to think that a "final cause" is necessarily a purpose. A purpose is merely that form of final cause which is most familiar to our experience [...] we must understand by final causation that mode of bringing facts about according to which a general description of result is made to come about, quite irrespective of any compulsion for it to come about in this or that particular way; although the means may be adapted to the end. The general result may be brought about at one time in one way, and at another time in another way. Final causation does not determine in what particular way it is to be brought about, but only that the result shall have a certain general character. (CP 1.211)

In modern biology, purposefulness is accepted only at the level of the individual: every individual organism somehow struggles for its proper existence, even microorganisms are directed towards goals which in their case is to reach a source of nutrition. J. Monod (1972) proposed the term *teleonomy* to differentiate individual goal-orientation from final causation in the Aristotelian sense.

Evolution of Evolutionary Thinking

To summarize the preceding parts of this chapter, the effort was to briefly give the main points of the three types of evolution by Peirce, thus to introduce Peirce's evolutionary theory. This can be characterized as triadic. The main two contradictory evolutionary forces of chance and law are brought together thanks to the force of Habit. At this point of my argumentation I would like to extend the line of reasoning to intersections with Darwinian and Lamarckian theories, but also with other crucial evolutionary theories, mainly those of Henri Bergson. Lastly, I will comment on Jacques Monod's philosophical views on molecular biology and the genetic script. What all the aforementioned biological theories by Lamarck, Bergson, Darwin, Monod and Peirce have in common is the principle of relation between two contradictory notions, these being chance and law.

The balancing between the two opposite forces is still the building block of today's biology represented by the Modern Synthesis of Darwinism. In Darwin's case, after the discovery of the genetic laws, many of his ideas obtained experimental support, and therefore more relevance. The paradigm of Modern

Synthesis builds upon the Darwinian evolutionary theory and gives it even more scientific relevance thanks to the notion of gene and genetic inheritance, these providing explanations of how random variation and inheritance actually works at the biochemical level. But what about Peirce and his achievements in evolutionary studies? Can one say that Peirce's conception of evolution gained relevance as the field of genetics changed the scientific status of biology as a discipline? Surely not. No textbook of contemporary biology refers to Peirce, but there is something we can note. At first glance, Peirce's idea of evolution is closely related to the teleological viewpoint, and this understanding of life is in direct contradiction with the mechanical paradigm of modern science. Also, the rhetoric of Peirce does not invite scholars to consider his writings on evolution other than as philosophical writings for pleasure, but this would be misguided. According to Peirce, every relevant scientific method proceeds by abduction, or hypothesis. What Peirce taught us was, after all and most importantly, that being a scientist does not exclude being a poet. And this not in a separate kind of way, but somehow continually, without a possibility to determine where the boundary is, like the color of the line delimiting the boundary between the red and blue part of a surface in Peirce's example from *The Law of Mind*:

> The color of the parts of a surface at any finite distance from a point, has nothing to do with its color just at that point; and, in the parallel, the feeling at that any finite interval from the present has nothing to do with the present feeling, except vicariously (Peirce, The Law of Mind *in Bisanz* 2009, 91).

At this very line, a non-defined line, lies the point where a poet becomes a scientist or vice versa. In like manner, Morris Cohen defines the uniqueness of the spirit of Peirce in claiming he was at the same time a medieval scholar and modern scientist (at xxvii, in *Chance, Love and Logic*, Ketner).

To answer the question whether there exists such knowledge (from the field of genetics or any other modern biological science) providing experimental support that confirms Peirce's original understanding of life and evolution as Love, we can move towards the direction of epigenetics and the Extended Evolutionary Synthesis (EES). But before I develop this topic, I will comment on the intersections of Peirce with the principal biological theories of the last century.

Peirce and Darwin

Peirce does comment on Charles Darwin in some of his writings related to evolution, but not exclusively. Given the fact that the theory of random variation is related to Peirce's studies on chance and probability, the evolutionary theory is in the center of Peirce's thinking from the period of the last decade of the

19th century (Burks 1997). The main outcome of Peirce's comments on Darwin is a critical consideration of the principle of random variation and, consequently, the survival of the fittest.

> The idea of the survival of the fittest is insufficient to explain the process of evolution. This is precisely the complaint that Peirce raises against the Darwinian theory of evolution in Evolutionary Love, after he has declared himself to be a Darwinian in his former articles. (Acosta 2015: 38)

The notion of chance or random variation is so closely associated with Darwin's theory that is impossible to think about it in different terms. Nevertheless, the image we have about Darwin's theory today is rather that of the Modern Synthesis of Neo-Darwinism, which is a Darwinian idea of random variations completed by Mendel's laws of genetics. As it occurs very often with theories that have become notoriously famous, at the end of the day we are dealing with but interpretations of interpretations. As a consequence, the original message of the author of the given theory might be somehow changed. A. Rosenberg (Rosenberg and McShea 2008) points to a certain misunderstanding of the notion of chance within Darwinian theory by the Modern Synthesis. Darwin used the term random variation, as he argued, out of ignorance of the actual cause, not because of the non-existence of causation, as it was often misinterpreted.

> I have hitherto sometimes spoken as if the variations […] had been due to chance. This, of course, is a wholly incorrect expression, but it serves to acknowledge plainly our ignorance of the cause of each particular variation. (Darwin 1987: 102)

Moreover, as Rosenberg noticed, the randomness of variation does not reside in mere ignorance of the prior causes, but resides, additionally, in the independence of a variation from the factors that determine its final adaptation (preserving the varied trait).

> The theory requires that in every generation heritable traits vary to some degree, and that this variation is "random." […] The theory of natural selection does however rule out one cause of variation in heritable traits, namely a future cause in which new variation is guided by the needs of the individual who bears it. Indeed that is the major thrust of the word "random" in the phrase "random variation" in Darwin's theory. It is not that the appearance of a new trait is undetermined, that it is not fixed by prior causes. It is rather that the causes that fix it are independent of, unconnected with, the factors that determine its adaptedness. We say that variation is random "with respect to" adaptation. (Rosenberg and McShea 2008: 18)

Darwin even expressed a necessity, a must-be, of the causes in species variations:

> Whatever the cause may be of each slight difference in the offspring from their parents – and a cause for each must exist – it is the steady accumulation, through natural

> selection, of such differences, when beneficial to the individual, that gives rise to all the more important modifications of structure, by which the innumerable beings on the face of this earth are enabled to struggle with each other, and the best adapted to survive. (Darwin 1987: 131)

The variation that has occurred is due to a certain cause in Darwin's theory, not random in connection with prior causes. Knowing the cause, however, of the modification is irrelevant to its effect, such as a better adaptation to the environment (or worse). We can therefore speak of randomness, of the chance of the variation with connection to its future consequences (not in connection with its prior causes). Hypothetically, other variations could have occurred to solve the situation with the same success. This is probably what is meant by the term chance in Darwin. The randomness does not mean, however, that there is not a causal relationship, it only means that the consequence of the variation (adaptation) is not necessarily linked to the very cause of the variation.

The aforementioned understanding of the principle of random variation is related to the functional plasticity of organs. The polyfunctionality of organs and body shapes was discussed by Darwin in a detailed manner. He also detected a relationship between the particularity of shapes and their specific functions and variability itself: the more an organ is particular, the more concrete function it has, the less it tends to vary:

> As long as the same part has to perform diversified work, we can perhaps see why it should remain variable, that is, why natural selection should have preserved or rejected each little deviation of form less carefully than when the part has to serve for one special purpose alone, in the same way that a knife which has to cut all sorts of things may be of almost any shape, whilst a tool for some particular object had better be of some particular shape. (Darwin 1987: 115—116)

Chance surely plays an important role in Darwin's theory, yet as we can see, Darwin also considered the environment, "conditions of life" having influence on variability. Additionally, the notion of chance is not so straightforward in Darwin; it is always related to the notion of natural selection. "Natural selection, it should never be forgotten, can act on each part of each being, solely through and for its advantage" (Darwin 1987: 116). From this formulation it is not clear that the advantage is for race rather than for an individual, but from the Modern Synthesis this aspect is more striking. After all, Darwin considered the principle of random variation in connection with the benefits for individuals:

> Whatever the cause may be of each slight difference in the offspring from their parents – and a cause for each must exist – it is the steady accumulation, through natural selection, of such differences, when beneficial to the individual. (Darwin 1987: 130)

Dawkins' theory of selfish genes (1976) represents an even more salient view on the opposition between individual organism and race. To conclude, it can be said that within today's reception of Lamarck, Darwin should be and often also is described as a Lamarckian, since he believed in the transmission of features from parents to offspring and in the role of the environment in heredity (Markoš and Švorcová 2019: 63). As Elliot points out, Lamarck would even consider himself as modern neo-Darwinian: "Had he lived in modern times, it is just as likely that he would have been a neo-Darwinian as a neo-Lamarckian" (Elliot in Lamarck 1963: XLVII).

In his paper from 1877, *The Fixation of Belief* (1877: 2–3), Peirce mentioned Darwin's name in connection with the use of statistical methods in science of biology. By this act, he fused two of the meanings of Chance: mathematical, law-like Chance as statistical method, and evolutionary Chance as Chance leading to Habit. Both meanings are present in the work of Darwin.

> Mr. Darwin proposed to apply the statistical method to biology. The same thing had been done in a widely different branch of science, the theory of gases. Though unable to say what the movements of any particular molecule of a gas would be on a certain hypothesis regarding the constitution of this class of bodies, Clausius and Maxwell were yet able, by the application of the doctrine of probabilities, to predict that in the long run such and such a proportion of the molecules would, under given circumstances, acquire such and such velocities ... In like manner, Darwin, while unable to say what the operation of variation and natural selection in any individual case will be, demonstrates that in the long run they will adapt animals to their circumstances. Whether or not existing animal forms are due to such action, or what position the theory ought to take, forms the subject of a discussion in which questions of fact and questions of logic are curiously interlaced. (Peirce 1877: 2–3)

Peirce and Lamarck

In *The Architecture of Theories* Peirce concludes that, while in Darwin evolutionary theory is centered around the striving of races, in Lamarck's case striving concerns individuals. Further, in the first case evolution is driven by Chance and in the second case evolution is driven mostly by Law. Even if critical, Peirce does not exclude either of them; on the contrary, he agrees with both of them, complementing both types of evolution by chance and by law with a third type of evolution.

The third type of evolution, the evolution by habit or love, is closer to Lamarck's evolutionary project. Peirce takes distance from both Darwin and Lamarck, yet he seems to have some more sympathy with Lamarck, at some places even comparing Lamarckian evolution with his proper conception of agapastic evolution:

> Evolution by sporting and evolution by mechanical necessity are conceptions warring
> against one another. Lamarckian evolution is thus evolution by the force of habit [...]
> Thus, habit plays a double part; it serves to establish the new features, and also to bring
> them into harmony with the general morphology and function of the animals and plants
> to which they belong. But if the reader will now kindly give himself the trouble of turn-
> ing back a page or two, he will see that this account of Lamarckian evolution coincides
> with the general description of the action of love, to which, I suppose, he yielded his
> assent. (Peirce, Evolutionary Love, in Bisanz 2009, 124)

Peirce found the essence of Love in the writings of Lamarck, even if Lamarck
himself didn't really develop interest in this direction. It is true that Lamarck
abundantly made use of the word "habit" in *Zoological Philosophy*, yet as Peirce
himself doubted, probably Lamarck's use of the term "habit" was not in the sense
of Evolutionary Love nor in a teleological sense. Let me repeat the quote from
the previous section to see in what circumstances Lamarck placed this term and
what he really meant by it.

> Nevertheless some of these herbivorous animals, especially the ruminants, are inces-
> santly exposed to the attacks of carnivorous animals in the desert countries that they
> inhabit, and they can only find safety in headlong flight. Necessity has in these cases
> forced them to exert themselves in swift running, and from this *habit* their body has
> become more slender and their legs much finer; instances are furnished by the ante-
> lopes, gazelles, etc. [...] It is interesting to observe the result of habit in the peculiar
> shape and size of the giraffe (Camelo-pardalis): this animal, the largest of the mammals,
> is known to live in the interior of Africa in places where the soil is nearly always arid and
> barren, so that it is obliged to browse on the leaves of trees and to make constant efforts
> to reach them. From this *habit* long maintained in all its race, it has resulted that the
> animal's fore legs have become longer than its hind legs, and that its neck is lengthened
> to such a degree that the giraffe, without standing up on its hind legs, attains a height of
> six meters. (Lamarck 1830: 122)[12]

Habit explains the result of evolutionary process: thanks to a habit of prolongation of the neck, giraffes have a lengthened neck. But this habit is not goal-
oriented, and in this sense is blind: habit is but a response to necessity, whether
it be physical constraints or environmental conditions. In words by Švorcová,
Lamarck "explained the variability of organisms on a behavioral basis as part of
the law of use and disuse" (Markoš and Švorcová 2019: 64). Yet Peirce stresses
that behind Lamarck's own words there is something hidden which he connects
to his theory of Evolutionary Love.

---

12   Italics inserted by author of the book.

Lamarckians further suppose that although some of the modifications of form so trans-mitted were originally due to mechanical causes, yet the chief factors of their first pro-duction were the straining of endeavor and the overgrowth superinduced by exercise, together with the opposite actions. Now, endeavor, since directed toward an end, is essentially psychical even though it be sometimes unconscious; and the growth due to exercise, as I argued in my last paper, follows a law of a character quite contrary to that of mechanics. Lamarckian evolution is thus evolution is by the force of habit. (Peirce, Evolutionary Love, *in Bisanz 2009, 124–125*)

Peirce associated Lamarck with evolution by law (as was mentioned previously in the section on Law), yet he also noticed a parallel with evolution by Love or Habit. We can see a parallel to the movement of Neo-Lamarckism, also because of the limits of the scientific knowledge of the period in which he lived.

Lamarckism is in today's interpretations associated with vitalistic or teleo-logical accounts, thus presumably contrary to the principle of evolution by Law. But originally evolution as inheritance of acquired features by Lamarck is not a vitalistic concept; on the contrary, inheritance of acquired features is a typical example of logical causality. Therefore, Peirce was right in claiming Lamarckism matches evolution by Law. The only reason to see Lamarckism as teleological or vitalistic is in the understanding of organisms as individual agents, not just passive vehicles of genetic information. Lamarck spoke about a natural power directing organisms to adapt themselves to the environment. Subjective teleol-ogy was nevertheless wrongly attributed to Lamarck by his successors.

Because of the postulation of such power ('pouvoir de la nature'), Lamarck has often been interpreted as a vitalist, though he considered such force to be physical (e.g., heat, electricity, magnetism). (Markoš and Švorcová 2019: 65)

This is probably why some authors (Otis 1994) merge the two notions of Lamarckism and teleology. The only acceptable notion in the context of modern scientific discourse is the notion of teleonomy (a term proposed by Monod, to be distinguished from teleology), which covers the individual agency of an organ-ism but not final cause in the Aristotelian sense. It has to be said that neither Darwin or Lamarck would probably agree with the modern interpretations of their theories.

Contrarily to Lamarck, Peirce was convinced that mechanical necessity is not the only force in evolution (and he also was convinced Lamarck himself expressed ideas of this kind). Or better, Peirce believed that there are rules in nature which are not mechanical, not clock-like, but rather like clouds. Karl Popper formulated in *Of Clocks and Clouds* (1966) a beautiful metaphor about the possibility of being scientific but not mechanistic:

So far as I know Peirce was the first post-Newtonian physicist and philosopher who thus dared to adopt the view that to some degree all clocks are clouds, or in other words that only clouds exist, though clouds of very different degrees of cloudiness... I further believe that Peirce was right in holding that this view was compatible with the classical physics of Newton. I believe that this view is even more clearly compatible with Einstein's (special) relativity theory. (Scott 2006: 64)

In connection to today's applications of Lamarck's ideas, I will briefly comment on what is intended by EES. In recent years, the very notion of gene itself has been freshly discussed in the field of genetics. Scherrer and Jost (2007) point out that the notion of gene is not particularly clear in that the gene as a script of a certain organic function or structure is in many cases not fully present at the level of DNA. Instead, the regulatory aspects of DNA are as important as the information itself encoded in the genetic script. With this point of view, the definition of gene and function as encoded in the nucleic base strings is unsatisfactory.

Specifically, the question „how much of the (human) genome has an identifiable function" is discussed controversially. Estimates range roughly from 5 to 90 % [...] Such divergence cannot be reconciled by more accurate data. Rather it reflects dramatic disagreements about the proper definition of function. (Laubichler et al. 2015: 144)

Epigenetic modifications, epigenetic inheritance, gene expression regulation and many other recently discovered molecular processes may reopen the discussion about what actually determines the organic structures and functions. EES encompasses all these processes participating in gene expression but which the DNA-centric genetic theory of the Modern Synthesis cannot abide.

The new discoveries argue that not only the virtual script, but also its context-dependent reading, determines the final result (cell, protein or organism) and that these expressions of the script can change due to environmental conditions, and the changes can be reversible, but in some cases can also be heritable. The inheritance of the "subjective" reading of the genetic script by an organism is obviously associated with Lamarck's idea of the inheritance of acquired features. Therefore, the movement of Neo-Lamarckism has gained some more support in the last few decades. There are nevertheless several inconsistencies between Lamarck's original theory and the movement of Neo-Lamarckism. In the words of Švorcová and Kleisner:

when speaking of a Lamarckian dimension, we do not refer to any sort of 'intentional' changes in genetic information, which is a claim frequently, yet erroneously, ascribed to Lamarck. Such anachronistic interpretations of his ideas – interpretations which historians would see as examples of presentism, that is, of understanding the past through the prism of contemporary perspectives (such as the discovery of concrete units of heredity) – are still widely spread and sometimes find their way even into academic writings (O'Leary 2015; Skinner 2016; Mastinu 2015). (Švorcová and Kleisner 2018: 233)

There is already considerable empirical evidence of the possible heritable epi-mutations of the genetic script or the ambiguity and multiple context-dependent interpretations of the reading of the genetic script (see more about epigenetics in the next chapter). The research is based on demonstrating several epigene-tic processes, e.g. how non-genetic inheritance is possible, how gene-regulation functions, how the external factors (environment) influence gene expression and what is the exact role of non-coding sequences in DNA macromolecules. Among these experiments, research in lateral (horizontal) gene transfer also appears to have an important influence on discrediting "the dogma" of the Modern Synthesis: lateral gene transfer, present mostly in bacteria, is a mechanism of sharing genes among one and the same generation[13]. According to the Extended Evolutionary Synthesis there is, apart from the DNA, a vast range of consider-ations that are needed to fully explain processes in living creatures, in order to reach a deeper understanding of the evolution of species.

Peirce and Bergson

The Darwinian theory of random variation (and thus the official scientific the-ory) had already existed for more than thirty years when Peirce entered the dis-course of evolution (*Evolutionary Love* is from 1893 while *Origin of Species* dates back to 1859). Thirty years is no small amount of time and it has to be said that in the last twenty years of the 19th century something like an official interpreta-tion of Darwin's theory already existed, along with its socio-economical appli-cations and connotations. The use of Darwin's theory for excusing or justifying the inequities of free market economics has subsequently become commonplace. Peirce seems to be familiar with this very interpretation of Darwin's theory, and it appears that the politico-economical application of Darwin is even more disturbing for him than the evolutionary theory itself. He uses the expression "Gospel of Greed" (Bisanz 2009: 122). At some point, Peirce even reverses the whole argument, saying that "Darwin merely extends politico-economical views of progress to the entire realm of animal and vegetable life"[14]. The role of chance in the evolution of species is not in itself disturbing for Peirce; he even agrees with chance as one of the rudimentary principles of evolution of the cosmos in

---

13   About the dogma and illusions of the Modern Synthesis see also Noble (2021), Bolshoy and Lacková (2021).

14   As Brier (2019: 67) notices, referring to Loye, "classical economic thinking and lib-eralism's development into social Darwinism has only picked the individual fight for survival competition aspect out of Darwin's much more holistic idea."

general. Peirce presents Darwinian and Lamarckian theories, as the vast majority of interpreters do, as opposite ways of understanding evolution, but he concludes by fusing them together and completing them with his own third type of
evolution, to acquire the triadic evolutionary theory based on both chance and
law, completed by something like the creative force of attraction or habit. Now
the question arises whether Peirce was the first to propose such an evolutionary
theory and whether his work is original in this sense.

The third mode of Peirce's evolution is Love, which is agapasm; some authors
understand it as "creative love" (Acosta 2015), others as a "law of mind to take
habits" (Brier 2006). If we decide for the moment to understand it in the sense
of creative love, we can see that the idea of creativity is shared by both Peirce
and French philosopher Henri Bergson. Creativity in Bergson's evolutionary
theory is related to the central notion of *élan vital*. *Élan vital* is the force of living beings to grow, live and fit to the world. This force is neither teleological
nor mechanistic-causal. The only characteristic we can attribute to élan vital is
creativity. According to Bergson, the living is irreducible to physical laws, and
the only possibility of reduction is the reduction to *élan vital*, which is by itself
impossible to explain by any kind of scientific language.

These two theories by Peirce and by Bergson emerged more or less in the same
period of time, even if Peirce's ideas came to light probably earlier than those of
Bergson: *Evolutionary Love* is from 1893 while *Creative Evolution* is from 1907.
Very probably Bergson was not familiar with the writings of Peirce. Might it be
what Peirce calls "the spirit of an age" (Peirce, *Evolutionary Love* in Bisanz 2009,
129), talking about a shared mind, a mind shared by several tendencies living
in the same period of time? In this case the idea of creative evolution would be
shared by Peirce and Bergson and would be a marvelous example of agapasm
and synechism of mind as Peirce describes in *Evolutionary Love*. At the end of
the day, this is also a demonstration of the agapastic way in which minds are
attracted one to another.

> I believe that all the greatest achievements of mind have been beyond of powers of
> unaided individuals, and I find, apart of the support of this opinion receives from syn-
> echistic considerations, and from the purposive character of many great movements,
> direct reason for so thinking in the sublimity of the ideas and in their occurring simul-
> taneously and independently to a number of individuals of no extraordinary general
> power. (Peirce, Evolutionary Love in Bisanz 2009, 133)

And still, Peirce's agapastic evolution is somehow distant from Bergson's
creative evolution. In the original introduction to 1998 Bison Books edition of
Chance, Love and Logic (ed. Ketner), Morris Cohen (xxv–xxvi) mentioned the
apparent similarity between Peirce and Bergson. Cohen explains why Peirce's

evolutionary theory is unique and in what ways it differs from Bergson's creative evolution. First of all, the most striking difference between Peirce and Bergson lies in that Bergson assumes no place for mechanistic explanations in evolution, but this is not the case of Peirce. Quite on the contrary, Peirce does not exclude mechanistic explanations from his theory; he is convinced that mechanistic explanations are not sufficient for explaining life and evolution, but this does not mean that they are not part of nature. Rather, mechanistic laws are to be completed by laws of habit taking. They are only one part of the threefold evolution for Peirce. Moreover, Cohen continues

> Instead of postulating with Spencer and Bergson a continuous growth of diversity, Peirce allows for growth of habits both in diversity and in uniformity [...] The creative evolution of Bergson though intended to support the claims of spontaneity is still like the Spencerian in assuming all evolution as proceeding from the simple to the complex. Peirce allows for diversity and specificity as part of the original character or endowment of things, which in the course of time may increase in some respect and diminish in others. Mind acquires the habit both of taking on, and also of laying aside, habits. Evolution may thus lead to homogeneity or uniformity as well as to greater heterogeneity. (Cohen, in Ketner 1998: XXVI)

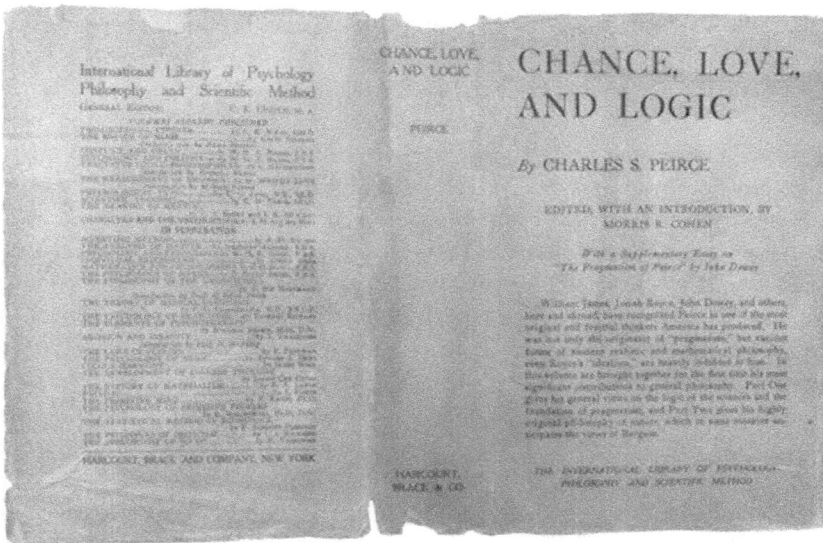

Figure 1: First edition, Chance, Love, and Logic, edited by Morris R. Cohen, 1923, picture taken from Peirce Institute for Studies in Pragmaticism, Texas Tech University in Lubbock, Texas.

Bergson's theoretical-philosophical approach to evolution is apparently the clos-est to Peirce's agapasm. Both theories operate with creativity in the evolution-ary process, yet fundamental incongruities are found between the two theories, and they are well expressed by Cohen. The main difference lies in the unlimited creative growth in Bergson, contrasting with the force of habit in Peirce: Habit does not always lead to unlimited diversity, but rather habit leads to a dynamical balance between chance and law, between homogeneity and heterogeneity. For more reading on Peirce and creativity in evolution see also Burks (1997).

Peirce and Monod

As was already mentioned previously at several points, the factor of chance was the most appreciated by the Modern Synthesis of Darwinism and still continues to be one of the major building blocks of contemporary genetics. Chance in the shape of random genetic mutations is familiar to every student who has taken an introductory class of genetics. But what about the notion of law? Laws also exist in the modern interpretation of Darwinism, by virtue of *code* – in the genetic code, for instance, law is due to evolutionary convenience or consummation of the energetic minimum. The code used for the information stored in the macro-molecules of DNA represents an evolutionary invariant: necessity. Nevertheless, the relationship between two elements consisting of the code (amino acids and DNA bases) remains arbitrary, because the relationship between the two is not guaranteed by any physical or chemical necessity, it having been, originally, optional. Indeed, if the relationship was originally arbitrary, it does not mean that it also continues to be optional after the establishment of the code. Once the code is established, its rules should be preserved. If the rules of the genetic code are not preserved, anomalies and pathological cases are the result. Both arbi-trariness and conventionality guarantee the existence of a code. Jacques Monod (Monod 1972) named the two inevitable parts of molecular biology as chance (*le hasard*) and necessity (*la nécessité*), chance referring to the optional character of the relation between the two parts of the code, necessity referring to the conven-tionality of the code. Once the rules have been established, they should be con-served, because in the inverse case, the code could not be used to communicate and to pass on information to the following generations. Monod (1972: 90) even used the term "arbitrary" when referring to the function of allosteric enzymes, defining the function of enzymes as being chemically arbitrary (*chimiquement arbitraire*), but also inventing a new term to design this arbitrariness. He named the relationship between the function of a protein (enzyme) and its chemical design as a relation of gratuity (*gratuité*). A. Markoš (2008), in the introductory essay to the Czech translation of Monod's book, noted that gratuity is a some-what peculiar term, since in French, it means both "for free" and "unjustified", but in English and in Czech the term tends to be understood as "for free". Markoš claims, however, that Monod's term should be interpreted preferably as "unjus-tified", due to the fact that from an evolutionary standpoint, it is not right to to say "for free", but on the contrary, the price was included in terms of how much energy was spent by previous generations. Thus, the synchronic "for free" state of chemical gratuity is counter-weighted by the diachronic energy consumed during the evolutionary process[15].

> The trait seems to be for free from a synchronic viewpoint, i.e. from a viewpoint of living organisms. These no longer need to take into account the long ages of evolution, during which the enormous amount of energy – both physical and creative – was invested in the improvement of each trait. (Markoš 2008: 177–178)[16]

We can take Monod's message as an illustration of the modern scientific approach towards evolutionary forces of chance and law. The relation between the necessity of DNA and the chance of evolutionary random variations com-prises the elementary principles of modern biology. Chance is still present mostly in the comprehension of evolution, but necessity is the stronger prin-ciple, with all the experimental support from molecular biology of the DNA macromolecule. As Burks notes:

> There has been a tremendous increase in our knowledge of biological evolution since Peirce postulated agapism to explain evolutionary creativity. Present knowledge consti-tutes a very strong case for a naturalistic or logico-mechanical explanation of evolution-ary creativity. (Burks 1997: 506)

Additionally, according to Burks (1997), Peirce's ideas about final causes in connection to evolution have been proved to be "wrong" (1997: 512). We have already demonstrated that actually it was not Lamarck but others including Peirce opting for teleological views on evolution. Now the idea of the direction I would like to take here is to somehow minimize Peirce's "wrongness", with the help of more recent research in biology. Burks' paper is from 1997, and since that time there has been major progress in epigenetics (Markoš and Švorcová 2019) and EES (Lacková 2018), in general proving scientifically and experimentally

---

15 Markoš focuses more on the term "gratuity" in the book *Evoluční tápání* (Markoš 2016).

16 In the original: "Zadarmo se vlastnost zdá být z hlediska synchronního, tj. Z hlediska právě žijících organismů. Ty už nepotřebují brát v úvahu dlouhé věky evoluce, během nichž se do vylepšování každé vlastnosti investovalo obrovské množství energie – fyzikální i té tvůrčí."

that organisms might inherit features acquired during the lives of their parents; it has been proven that direct causality plays some role in heredity. But when it comes to the notion of teleology, the question becomes more delicate and we should also approach this in such a delicate manner. In accordance with evo-devo studies, teleonomy is a better candidate for the lexicon one should carefully selected. Because the term of teleology is really difficult to prove experimentally and is in contradiction with the current scientific paradigm, it is avoidable. But this does not mean that we cannot study Peirce's agapastic evolution from an up-to-date viewpoint, with all the enormous progress in biology that was made in the meantime.

Not only chance and necessity, but the triadic understanding of evolution is the only way Peirce could encompass the notion of life and only in the triadic manner can EES become a new way of introducing the semiotic understanding of life.

Using the definitions of terms such as chance and necessity (Monod), random variation and preservation (Darwin), or chance and law (Peirce), we can see an evident continuity of thinking among the evolutionary theories not only at the level of evolution of species (Darwin and Peirce), but also at the level of molecular biology and the development of the genetic code (Monod). This continuity resides in understanding life dynamically as a tension or balance between what is arbitrary and what is conventional, between chance and law. The innovation of Peirce might be found in that his evolutionary theory is not limited to the dualistic conception of living phenomena. He introduces agapastic evolution as the third component of evolutionary dynamics. Peirce's definition of life is triadic and therefore semeiotic. Peirce's evolutionary theory did not receive much consideration from evolutionary biology. In the words by N. Houser, "Peirce's synechism and agapasm have yet to be discovered by modern science" (Houser 2014: 29). Pietarinen (2011) for example proposes that the so-called Baldwin effect (influence of learned behavior on the genome of the species) is in total accordance with Peirce's agape and evolutionary love. I believe that current biology, especially epigenetics and protein studies, is the most promising field of application for Peirce's theory.

Chapter 3   Peirce's Relational Realism: Real Possibilities and Epigenetics

Introduction to Epigenetics

Peirce's triadic understanding of evolution completes his views on biology as a scientific discipline as it was established between the nineteenth and twentieth centuries, but Peirce's role in biology as a scientific discipline does not end with his papers dedicated to evolution. Contemporary biology might also benefit from Peirce's relational and triadic patterns, and the need for a triadic logical understanding of molecular processes becomes even more striking with the discovery of epigenetic processes and other recent disciplines appearing after the 1970s. In an attempt to demonstrate the universality and uniqueness of Peirce's scientific method even in the context of the newest scientific discoveries, I will go through the current research in epigenetics and show how the current state of biological research is no longer describable in terms of dyadic models. Since Lamarck, Darwin and Peirce, biology has undergone an enormous paradigm shift with deciphering the genetic code and the related importance of the field of genetics. I already mentioned some similarities in the thought of Peirce and genetic insights on biology by J. Monod, but the current state of investigations in biology goes beyond genetics, also into the so-called post-genomic era. In the natural and human sciences, these new approaches, often summarized under the name of *epigenetics*, have been gaining popularity in the last decades while in the general public understanding the gene-centric position still prevails.

Various biological disciplines contest the central role of the genetic script, for instance evo-devo studies, niche construction theory (related to the notion of ecological inheritance), and horizontal gene transfer. But the word used most often is *epigenetics*, already mentioned in the previous chapter, sometimes also misused, or used with such a wide range of meanings that it is difficult to understand what the notion refers to. In very simple words, it can be said that epigenetic discoveries, in every possible synchronic sense of the word, contradict the central dogma of the modern synthesis of Darwinism, that is, the unidirectionality of determination going from genes to proteins (DNA $\rightarrow$ RNA $\rightarrow$ protein). According to the encyclopedia definition (Allis et al. 2015: 50) the central dogma is related to three crucial points: (1) "self-propagation of DNA by semiconservative replication" (2) "transcription in unidirectional direction" from 5´to 3´, and (3) protein creation unidirectionality from RNA to protein. In (1) the word semiconservative is used because actually RNA can also influence DNA, while in (3) the order is strictly unidirectional. I will explain in more detail the whole process of protein synthesis in Chapter 4.

The word epigenetic became famous thanks to Conrad Hal Waddington, one of the founders of systems biology and evo-devo biology. The original meaning of the word epigenetics was related to the development and differentiation of tissues, cells or organs of an organism: "determination, an idea which in effect implies that at some early stage in development certain major discontinuities between tissues or organs become established" (Waddington 1957: 14). The current use of the word *epigenetics* is more often associated with concrete molecular processes happening at the genetic level but not having direct influence on the very genetic script, but it also has many other meanings, more generally expressing the complex relations between genotype and phenotype.

To illustrate epigenetic phenomena, Waddington used a visual metaphor – the epigenetic landscape – to depict the developmental patterns of embryogenesis. The epigenetic landscape describes a ball rolling down a hill to an inclined landscape, with various terrain containing many hills and valleys. The model metaphorically represents the initial state of the fertilized egg and its ontogenetic development adapting to the environment during the different developmental stages.

Even if all the meanings of the word epigenetics share the rejection of the central dogma, they also exhibit very different features. What is the most attractive about epigenetics even for the humanities and ending up in applications such as cultural epigenetics (see more in the next section of this chapter), is the non-dogmatic approach. This approach has also been disseminated as a somewhat more open reading or interpretation of the strict and frozen understanding of the genetic script. In this optics, the genetic alphabet, composed of the four basic letters (or bases) A, C, G and T (adenine, cytosine, guanine, and thymine), formulates "words" – genes; so if "genetics" equals words, "epigenetics" equals the reading of these words (Allis et al. 2015: 48). This is a very broad understanding of the term epigenetics, and also to some extent imprecise.

> Consider a more or less flat, or rather undulating, surface, which is tilted so that points representing later states are lower than those representing earlier ones. Then if some thing, such as a ball, were placed on the surface it would run down towards some final end state at the bottom edge. There are, of course, not enough dimensions available along the bottom edge to specify all the components in these end states, but we can, very diagrammatically, mark along it one position to correspond, say, to the eye, and another to the brain, a third to the spinal cord, and so on for each type of tissue or organ. Similarly, along the top edge we can suppose that the points represent different

cytoplasmic states in the various parts of the egg. Or we could represent the various different initial conditions by imagining various degrees of bias on the balls which are to run across the surface. (Waddington 1957: 29)

Epigenetics accounts for features that, if considering only and exclusively four letters of the genetic alphabet, would remain unexpressed: epigenetic modifications can be chemically expressed by the addition of a special mark, for example, adding a methyl group to the DNA molecule so that one of the four letters of the genetic alphabet becomes a marked letter, for instance cytosine becomes methylated cytosine, a modification which serves many cell processes such as aging or inactivation of the X chromosome.

The establishment of the genetic database known as Encode was a historic moment in genetics (Consortium 2012). Encode is the largest genetic bank for the human genome and the first systematic attempt to work out what is the role of the so-called *junk DNA*. It was known already decades ago that less than 2 % of DNA codes for proteins. The rest of the genome was considered as historical residue, "junk", unnecessary segments. Thanks to the Encode project it was revealed that in fact, the believed "junk" DNA does encode important biological functions. The "junk" DNA does not encode proteins directly, but it does encode processes helping the proteins to be synthesized and regulates the whole process of protein synthesis. They found that about 18 % of our DNA sequence is involved in regulating the gene expression and about 80 % of the DNA sequence can be assigned some sort of biochemical function.

Non-coding RNAs (ncRNA) are a transcript of non-coding DNA. The existence of abundant noncoding RNA with essential biological roles (e.g., transfer and ribosomal RNA) has been known now for a long time (Tollefsbol 2019: 308). To study non-coding RNA, the NONCODE project was established. Experimental evidence has shown that long non-coding RNA (lncRNA) "plays important roles in several biological processes including transcription, splicing, translation, cell cycle, imprinting, pluripotency, and dosage compensation. Moreover, lncRNA are implicated in numerous pathologies, including cancer" (Tollefsbol 2019: 308). Moreover, it has been proposed, and evidence proven, a role for lncRNA in human and mice germ line development (Tollefsbol 2019: 332).

Several studies on DNA methylation have been conducted concerning maternal care in rodents (Weaver et al. 2004; Cameron et al. 2008). In these studies, maternal rat licking and grooming was identified as an important factor in the adult sexual behavior of the offspring. Two kinds of rat mothers were studied, high-LG (licking and grooming) mothers and low-LG mothers. The results suggest that female offspring of low-LG mothers show an increase in sexual receptivity. These variations in rat behavior are not random and are observable at the chemical level by the DNA methylation, which functions as "diacritical" changes at the DNA script. The DNA methylation can be caused by many external factors, as in the indicated studies, it can be behavior of the first generation, but it can also be the influence of the environment, such as temperature, pressure, nutrition etc. The externally caused epigenetic marks on the genetic script may become hereditary or transgenerationally inherited, thus transmitted to other generation(s). Moreover, they can remain only prenatal or postnatal, meaning that they are transmitted only and exclusively to the next generation (from parents to offspring) and not beyond. Epigenetic inheritance can bring populations to different genetic and phenotypic equilibria. This is a crucial observation for the upcoming considerations of epigenetics and semeiotic grounding of epigenetic phenomena.

Epigenetics in a narrow sense is thus nothing more than another set of letters added to the original four-letter genetic alphabet, thus, "just" another biochemical process taking place within the genome. Of course, epigenetic marks do not obey strict and immutable laws, as is the case of the genetic alphabet. Epigenetic marks are not always present, and their presence might be a reaction to some external factors, thus non-intrinsic. But it is interesting to note in the previous quotation by Tollefsbol the use of the expression "epigenetic mechanisms".

> Many definitions of heredity in contemporary biology, and in society at large, are explicitly genetic. Accordingly, evolutionary theory almost exclusively relies on transmission genetics to represent heredity (e.g., population genetics and quantitative genetics). But lately, the heredity concept appears to be changing [...] Under this definition, inheritance does not just refer to the passing of DNA from one generation to the next, but instead to parental transference of all the developmental means that enable reconstruction of the offspring developmental niche [...] Epigenetic mechanisms have received particular interest in this context. The aim here is to review our current understanding of the consequences of such epigenetic inheritance for evolution and how those mechanisms themselves evolve. (Tollefsbol 2019: 482)

In fact, epigenetic processes can be very well understood mechanistically. But as it was with evolution in Peirce, mechanism is a necessary part of the biological process. The important point is that mechanism exists in biology, but is not the only driver of biological processes. Probably every biological process is partially mechanistic, yet also arbitrary (chance) but most of all, also habitual. What is so interesting about epigenetics is that the habitual part is the strongest among the three components (chance, law, logic); but in order to retain a correct analysis, we should not omit the mechanistic part of epigenetic processes, and in order to understand what is going on at the molecular level when we talk about epigenetics, some introductory words for epigenetic processes are needed. Waddington's original definition is not that far from the current understanding of the term epigenetics, the difference being that the original idea of the processes of differentiation of cells from one eventually multifunctional and undifferentiated cell (germ cell or egg cell, as Peirce understood it, see previous chapter) into concrete molecular patterns, were experimentally discovered and classified as epigenetic:

> All cells in the human body originate from the totipotent zygote, and the instructions for differentiation of the specialized cell types are in part encoded by epigenetic signatures including DNA methylation, histone modifications, and small RNAs. Most of our knowledge surrounding postfertilization epigenetic reprogramming in the embryo are insights from mouse [...] In humans, epimutations at imprinted genomic loci are involved in the dysregulation of placental growth and the development of disease later in life. (Tollefsbol 2019: 198)

It should be noted that epigenetic processes are not the only molecular processes associated with disclaiming the central dogma of the Modern Synthesis and the possible plurality of interpretations of the genetic script. We can also talk about gene expression, gene regulation, phenotypical and developmental plasticity.

Phenotypical plasticity influences the reading of the genotype by the environment, leading to some modification at the level of phenotype. The modifications might be behavioral, morphological (bodily structures), or physiological. Polyphenic traits result which are multiple phenotypes originating from one single genotype. A notorious example of phenotypical modifications are the particular butterfly wing color-patterns produced physiologically in response to environmental stress, such as temperature conditions (Hiyama, Taira, and Otaki 2012). As with many other epigenetic mutations, the particular wing colors and patterns can become heritable and transmitted to the next generation. Pigliucci et al. (2006) define phenotypic plasticity as "the ability of individual genotypes to produce different phenotypes when exposed to different environmental conditions".

Developmental plasticity on the other hand reflects modifications of an organism during the life course, and these might as well be explained in terms of external environmental factors. Developmental plasticity also occurs, similarly to phenotypical plasticity, through gene–environment interactions and leads to morphological, physiological or behavioral changes. It might sound banal and obvious, but we often overvalue the importance of the genetic script, thinking that we are, as biological organisms, determined by genes. Think about what influence the diet has on our bodies, behavior and mental states and what enormous modification one organism can undertake by modifying its diet. Developmental plasticity refers to the individual ability of an organism to modify its development in response to environmental conditions (Moczek et al., 2011). Under these conditions, developmental plasticity makes it possible for alternative traits to express new environmental adaptations. Adaptations may even be transmitted to the genome through a process of genetic accommodation, leading to phenotype-genotype reciprocal correlation.

Phenotypical and developmental plasticity may correspond better to the original definition of epigenetics by Waddington and the early understanding of the term. An interesting account of epigenetics can be found in a book about structuralism by Swiss developmental psychologist Jean Piaget:

> Just as there still are embryologists who remain wedded to an entirely preformational view of ontogenesis and who, accordingly, deny all epigenesis (restored to its plain sense by Waddington), so it has occasionally been maintained of late that the entire evolutionary process is predetermined by the combination established by the constituents of the DNA molecules. Thought through to the end, preformational structuralism simply cancels the idea of evolution. Waddington, by reestablishing the role of the environment as setting "problems" to which genotypical variations are a response, gives evolution the dialectical character without which it would be the mere setting out of an utterly predestined plan whose gaps and imperfections are utterly inexplicable. These advances in contemporary biology are all the more valuable to structuralism because, joined to ethology [...] they furnish the basis of psychogenetic structuralism [...] Thus, the contacts with experience and fortuitous modifications due to environment on which empiricism modeled all learning do not become stabilized until and unless assimilated to structures, these structures need not to be innate, nor are they necessarily immutable, but they must be more settled and coherent than the mere groupings with which empirical knowledge begins. (Piaget 1970: 50–51)

Even though during the times of Piaget the empirical evidence for epigenetics was not so immense as it is today, his understanding of the term corresponds perfectly to current widened definition. It is noteworthy that Piaget does not deny the importance of structures in biology, only stresses that structures are not immutable and are potentially dynamic. This understanding of structuralism perfectly corresponds to one of the aims of this book: presenting diagrammatic models for biology, but with dynamic potential and without the prerequisite of causal pre-determinism or determinism in general. As will be demonstrated in the following parts of the book, "structuralist" thinking in biology need not necessarily point to the movement of biological structuralism from the last century. Peirce's existential graphs and Hjelmslev's participative logic represents a more suitable paradigm for the formalization of biological structures, conserving their immense potentialities and dynamic nature.

In their recent book about epigenetics, the authors say "life is primarily epigenetic" (Markoš and Švorcová 2019: 181), meaning that it cannot be explained from just one perspective, genetic or otherwise. Life is complex and dynamical, and must be approached as such by science. Only with a plurality of approaches can we define life or get closer to the definition of life. "Life is primarily epigenetic" means also that the main importance is given to the role of environment and the plurality of relations determining living organisms: parental, genetic, vertical but also horizontal.

It must be said that while we have considerable experimental evidence of molecular processes happening at the epigenetic level, their wider consequences and evolutionary importance remain yet to be discovered:

> Although epidemiological studies have been useful for studying the association of exposure and the transmission of health outcomes across generations, there is still a lack of understanding of epigenetic inheritance in humans. This is due to many factors including the population sizes and long follow-up periods required to complete this type of study. Additionally, accounting for differences in the frequency, magnitude, and duration of environmental exposures is difficult, and these factors affect epigenetic patterns. (Tollefsbol 2019: 199)

Texts, Scripts and Scriptures

In the illustrative examples of epigenetics, the metaphor of diacritical marks is used. It was already mentioned before how DNA methylation can be illustrated as a marked letter of the genetic alphabet, meaning that one of the four (or five) letters representing DNA bases (adenine, guanine, cytosine, thymine, uracil) becomes marked with the methylation mark, understood as a diacritical mark such as for instance C—C´. The metaphor works better for readers of languages with alphabets actually containing diacritical letters such as Czech, which is why the diacritical metaphor is very popular in texts by Czech biologist Anton Markoš. We can see in his own name, the final letter "š" is diacritically marked, referring to a distinct phoneme in the Czech language. Differently from s, š is pronounced as "sh". In general, the diacritical mark "ˇ" in Czech and some other Slavonic languages represents a palatalization of a consonant. The metaphor has its limits, of course, but it is only an extension of the decades old existing language or textual metaphor of the genetic script. The language metaphor of life started with the deciphering of the genetic code and the observation of the fact that it really resembles language as an arbitrary set of rules and correspondences between two distinct worlds connected only thanks to the arbitrary and conventional code. Thus, the genetic code was established by the scientific community as such because of having all the characteristics of a code. At that time, many linguists also joined the discussion about the linguistic nature of the genetic code, among them Roman Jakobson, famous member of the Prague Linguistic Circle, and who claimed a real existing analogy between DNA and natural language (Jakobson 1971; for more about the language metaphor of life see Markoš 2002). Jakobson established the analogy very precisely, according to his intuition and the information associated with the first published and popularized understanding of the genetic code, as follows: DNA and natural language share the same structure, and we can say that this structure follows the rules of double articulation, as André Martinet established, and the systems are accordingly stratified into letters, words and sentences or texts in the case of language, and into genetic bases, triplets of nucleic bases, and genes or genomes.[17]

The metaphor became sedimented and popularized at least at the level of analogy between letters and bases, so much so that even today we can find in many biology textbooks the DNA bases synonymized as "letters of the genetic alphabet". The analogy as established by Jakobson according to his linguistic intuition was recently questioned and tested experimentally on a number of genetic texts using methods from quantitative linguistics. It was found that according to mathematical textual laws present both at the linguistic and genetic levels, the analogy by Jakobson does not hold in an integrity (Faltýnek, Matlach and Lacková 2019): units at the level of letters (bases) should demonstrate the same mathematical-textual behavior (that is, demonstrate the same mathematical patterns according to Zpif's mathematical linguistic laws) in order for the analogy to hold. The experiment showed that, according to their mathematical behavior, bases do not correspond to language letters. Instead, DNA bases mathematically correspond to "distinctive features" in phonology. Distinctive features are the smallest existing units in linguistic analysis and correspond to speech sounds. The system of distinctive features is based on oppositions represented possibly as one and zero. In the Czech alphabet for instance, six such oppositions exist. Each letter of the Czech alphabet is represented by a unique string of six positions occupied by one or zero (accordingly, the present or absent arbitrary distinctive feature). In this study, written language was considered rather than spoken, mostly for the sake of simplicity of mathematical textual methods.

---

17   Roman Jakobson doesn't develop the analogy any further but Sungchul Ji for example
       takes it a few steps farther (Ji 1997).

Non-alphabetical languages, such as Chinese with its ideogrammatic characters instead of letters, were also considered for the study and this inclusion of Chinese characters into the study brought interesting results. Chinese was inserted to control the correctness of the method at the level of analogy between singular language units: words-letters-distinctive features.

The afore described experiment is important for noticing that Jakobson's original analogy was positioned too high. Instead of comparing genes with sentences and texts, we should compare genes with words, going one level lower in the analogy and starting the description not with letters but with distinctive features. The final message might be more significant than it seems at the first glance. The lowering of the position of the analogy points to the fact that our general understanding of genes and their immense (deterministic) power is diminished even mathematically. In the linguistic analogy, genes do not correspond to sentences and higher textual units, thus are not conveying meanings at the textual levels. They only convey meanings at the word-level, that is, they are incomplete semeioses. They are pure rhemas in Peirce's classification of signs, rhemas without predicates. I am using here Peirce's terminology of his classification of semeioses. According to the particular relation between R and I a semeiosis can be either rhema, dicent sign or argument.

> In regard to its relation to its signified interpretant, a sign is either a Rheme, a Dicent, or an Argument. This corresponds to the old division Term, Proposition, & Argument, modified so as to be applicable to signs generally. (SS: 33–34)

To become dicent signs, genes must be completed by other semeiosis components[18]. This is what epigenetics says in fact. A single gene is a pure rhema, and only when in the context of other genes, transcriptional and translational external factors, environmental factors, and with aid of chaperons and chaperonines, can it can potentially become a fully functional semeiosis with all the three components of object (O), representamen (R) and interprenat (I) and possibly classified as a dicisign or argument.

The questioning of Jakobson's metaphor brings us towards questioning the textual level of genetic phenomena. "Genetic text" is an expression almost as widespread as "letter of the genetic alphabet". The analogy of letters was proven to be wrong; accordingly, the validity of textual analogy should also be reexamined. "Genetic text" is a term used for a plurality of reasons. One has already

---

18   In the case of bacteria, the semeiosis model would be probably different: bacteria don't have introns and do not have an epigenetic aspect; one could assume that semeiosis is questionable in the case of bacteria. I leave this question open for experts in bacteria.

been mentioned, emerging from the linguistic metaphor and analogy of genetic bases with letters: the text metaphor is consequently only an extension of the letters metaphor. Another reason is related to the progress of bioinformatics in the last decades plus the fruitful usage of electronic gene databases, repositories of electronic transcripts of biologicals sequences. The electronic sequence transcription can contain DNA strings, RNA strings or other types of strings coding for the human genome or different types of organisms' genomes. It can be said that genome databases are biological counterparts of language corpora in linguistics (Bolshoy and Lacková 2021), both serving for a data-oriented quantitative studying of sequences, linguistic or biological with the aid of various types of programs or analytical software.

But we can ask whether these methods for textual analysis, both in linguistics and in biology, are powerful enough for an accurate and exhaustive description of a text, whether it be linguistic or biological. This question is related to the question *what is text*? There are many possible understandings of the word text, both in the restricted linguistic sense and in the widened general sense. The restricted linguistic sense of the word text includes linear sequences of words or other linear language units. It is not of particular importance if the text is written or spoken, since the criterion of linearity is conserved in both and in the written form linearity is dependent on spatial consecutive relations, whereas in the spoken form linearity is dependent on temporal consecutive relations. In both cases, linearity is inseparable from sequentiality and it is impossible to superimpose two units at the same time[19]. Linearity and the impossibility of superimposing two or more units at the same time are given by the physical constraints of language. Language is dependent on its material realization: even though linguistic meaning might be non-linear, the expression is necessarily linear. I argue in my dissertation (Lacková 2018) for the priority of the non-linear aspect of meaning over the linear aspect of the expression both in biology and linguistics, yet I also argue that one presupposes the other and we cannot separate meaning from expression. That is why pure quantitative methods dealing exclusively with

---

19   The only units that are actually superimposed and do not follow the rule of sequentiality are the distinctive features of a language. Yet it is questionable to what extent they can be comprehended as real linguistic semeiotic units, since they are neither direct (morphemes/words/lexemes) nor indirect (phonemes) bearers of meaning. In Hjelmslev's definition and terminology, they are neither signs nor figurae. Even though, at the molecular level, distinctive features are actually sequential, it suffices to look at a random spectrogram of a phoneme to experimentally demonstrate this observation.

sequences will never lead to satisfactory and exhaustive description of a text/language. Texts are linear, but only at the level of their expression. Texts are primarily bearers of meanings, and meanings are irreducible to linear, that is, dyadic relations.

In text linguistics, usually seven criteria of textuality are listed: cohesion, coherence, intentionality, acceptability, informativity, situationality and intertextuality. These criteria are the basics of text linguistics as formulated by De Beaugrande and Dressler (1981). If we observe these criteria carefully, we can see that they all touch the meaning and content level of the text. One may say, text criteria are mainly oriented towards semantics, some of them towards pragmatics, but none of them are oriented towards the expression-formal part of the text. Moreover, all of the criteria are pluridimensional, going in all directions from one text to another in the case of intertextuality, from text to extra-textual reality in the case of situationality or towards and backwards within the text itself in the case of cohesion and coherence. Consequently, the linearity criterion somehow disappears when we move from "lower" language levels to "higher" levels such as text. Thus, when we go back to the term "genetic text" and its implications, we have to keep in mind all of the problematic parts of the definition of text and textuality.

In the works of second-generation semiology, texts are understood as not necessarily verbal, yet any kind of semiotic system can be understood as a text and treated as such. An artwork is a text, a person is a text, a cultural event is a text, a meal is a text. This is not the dimension of the term "text" used in genetics and biology. In genetics and in genome databases, the term text is used in the more restricted sense of just a concatenation of sequences, which is per se a pleonasm. It would be more accurate to say that, from the restricted sense of the word "text", the text is but a sequence. It has no relevance whether we use blanks or other marks to demarcate zones of singular units, which is just an aesthetic or practical question of orthography, absolutely unnecessary from the functional perspective. Above all, in many writing systems demarcations of units do not exist (as it was in the case of old Latin). This is also the sense of the word *text* used in sequence biology, a field applying quantitative linguistic methods to biological sequences (Bolshoy et al. 2010). But in the field of epigenetics the restricted sense of the text is no longer satisfactory. Epigenetic "texts" are, in a way, also just long sequences, enriched with some unusual and temporary special diacritics. But epigenetic texts must be interpreted –consider the original meaning of epigenetics in the developmental and cell-differentiating sense, as nothing but the interpretation of the sequence, not the sequence itself. Moreover, epigenetic texts are themselves already interpretations of previous (epi)genetic texts.

We can comment also on the self-evident vocabulary in molecular biology referring to linear sequence comparison with written texts of human language. The vocabulary surely derives from the original language metaphor of DNA. Besides the word code, and the use of the expression "genetic letters", plenty of other linguistic terms are used in biology. The whole process of protein synthesis, where DNA is used for the production of proteins, is intertwined with linguistic vocabulary. Protein biosynthesis starts with DNA to RNA *transcription*, where the term primary *transcript* is used; afterwards, RNA is *translated* into amino acids, the so-called peptide chain which folds itself and results in the final protein; protein folding has been compared to a *syntax* in human languages, because of the relations between distant parts of the peptide chain brought together thanks to the fold; genetic *script* is used quite often, among the aforementioned terminology, and this term leads to the more concrete textual metaphor of *scripture*. Scriptures are holy texts of the world's religions, and in the context of biology it is meant to invoke profound associations with nature, life and self-reproduction. In the writings of Markoš the religious metaphor is not accidental. As he stresses, both (epi)genetic and holy texts are most intimately connected in the sense that the interpretation of the text matters more than the text itself. This statement is supported by the fact of how many different interpretations of one text exist (both in holy scriptures and genetic scripts) and what different lineages of cultures or organisms the interpretations lead to:

> Many different cultures are known that have developed from common roots and are based on identical or very similar generic, canonic texts. How many cultures have arisen based on different interpretations of a single canonical text – the Bible? We will find no difficulties here, because, along with the text, people (or peoples) also transmit the way to interpret it [...] But who is the interpreter in a biological species?
>
> In addition to the canonical text – two versions of genome inherited from our parents – we also inherit a small but very important piece of body: the egg cell itself. This is the agent that reads the dead text of genetic inscription and transforms it into information, technical documentation that can be consulted according to the situation. And this arrangement is the clue to the species-specific interpretation, that is, recognition of the text and of signals from outside on the basis of the history, experience of the cell, cell lineage, and species. The fertilized egg manipulates the genetic text according to rules inherited from countless generations of its predecessors. Driven by this tradition, it builds the specific morphology. I maintain that a species can be understood as a culture; it follows that the emergence of new species may equally be a matter of the mutation of the text (DNA) and/or changes in the rules for manipulating it. (Markoš 2002: 42).

The narrative aspect of holy scriptures is crucial for the epigenetic understanding of life. Narratives are continuously transmitted stories from generation to generation, changing over time, depending on who narrates and who interprets them. The difference between psalms and narratives lies in the relatively deliberate way of re-narrating the story by a who (that is, by a subject with a cognition). Epigenetics is based on the same narrative principles; every organism has possibility to renarrate but in the sense of interpretants that are non-whos. the story with slight modifications, or epigenetic marks. But there are several important questions we should ask when dealing with the narratological analogy between scripts and scriptures. The first question is *what* does the genetic story narrate? What is the story the genetic script conserves? The genetic script contains information about what we are, what morphological forms we are going to develop, what physiological characteristics we are going to have, etc. But as was already pointed out with the non-coding regions of DNA, the major part of the genetic script contains information with instructions for how, when, or when not to actually produce such and such morphological form. These "meta-data" are crucial and are also analogical to sacred scriptures containing instructions and concrete indications about *how to* perform religious rituals.

Another question related to the narrative nature of epigenetic scripts is *what is the function* of the storytelling by biomolecules? In holy scriptures the function is the cultural or spiritual passing of wisdom from one generation to another. Religious stories, myths and legends, explain the creation of the world, our place in it, and its nature. Besides this representational function they also function as some examples of how we should behave and live in this world. In epigenetic scripts the function is actually quite similar. Epigenetic marks are direct responses to the environment, and thus are the representations of the world and at the same time examples for future generations, when the epigenetic mark is inherited. This analogy leads to a peculiar question about what is the primary function of the genetic script: is its function communicative or representational? This question is an extension of the identical question about natural language (Chomsky, Berwick 2016). According to Faltýnek, the primary function both in language and in the genetic code is not communicative but representational (Lacková and Faltýnek 2021). The communicative function is only secondary to the primary need of representing the world we live in. In evolutionary biology, this question posited is somewhat differently, yet at the end of the day it is nothing but the very question of the primary function of the genetic code. The question, never answered, is about the origin of life on Earth. Some scholars are convinced of the "information first model", where the existence of the genetic code is presupposed in order to have life. This model brings some serious problems regarding how the genetic code could emerge from a total blank.

At this place we should also distinguish between the concepts of information and meaning. The usage of the term "information" varies a lot between authors and while some authors use it in wide way as a human communicative meaning, other authors use it as a purely cybernetic/informatics term (transmission of signals). To avoid any misunderstandings, it is better to distinguish information from meaning as suggested by Brier:

> But since the information concept is now firmly rooted in computer informatics and in the information theories of Shannon and Weaver as well of Wienerian cybernetics, another strategy would be to abandon the original human communicative meaning of the concept. This is what I will suggest as a strategy, because this will also make it possible to combine this theory with a semiotic sign concept without any major overlap between subject fields, thus paving the way for an integrative cybersemiotic framework instead of a paradigm competition. (Brier 2008b: 62)

In the context of the genetic studies, the term genetic information indeed usually designates this very understanding of the term "information" as different from "meaning". It is also why we have fields such as bioinformatics, because indeed we can treat genetic information exactly as we treat information in informatics. Yet in the paradigm of the EES, there has been a trend to switch from information to meaning and to understand evolution in the sense of interpreting meanings rather than replicating and copying information.

Returning to our previous discussion about the origin of life on Earth and the role of the genetic code in this process, in the "information first model" the term "information" is also used in the cybernetic way, to designate the information stored in the form of some proto-replicators (not necessarily DNA as we know it today). The early essays on the genetic code by its discoverers even hypothesize an extraterrestrial insertion of the genetic code on Earth: its structure is so complicated that it is improbable that it could arise by pure variation and natural selection. Other scholars on the other hand prefer "metabolism first" model. This model presupposes the beginning of life without replicators and the emergence of information storage and replicators only as further evolutionary steps. Numerous hypotheses about the emergence of prebiotic life have been proposed (Markoš and Švorcová 2019: 10–19; see also Teresa 2018). The problem of the metabolism-first model is that it does not explain how the emergence of protein synthesis is possible in terms of catalysts: in protein synthesis, it is proteins themselves which catalyze the chemical reactions responsible for protein biosynthesis, creating a hen-egg problem. A possible solution to this problem is the "RNA first model" of the beginning of life on Earth, according to which RNAs are polymers possessing both catalytic activity and information storage capacity. Consequently, RNA molecules might have been at the origin of everything, responsible for catalysis and storage at the same time. This model is often criticized for the fact that many laboratory tests have been made trying to synthesize such RNA molecules capable of self-replication and other needed molecular processes, yet so far none of them has been successful. Further alternatives have been proposed about polymers that resemble RNA but are chemically simpler, RNA-like polymers (Alberts et al. 2002).

All the alternatives to the "information-first model", together with the huge expansion of the field of epigenetics, go in the direction of decentralizing the role of DNA from molecular biology and genetics; more than that, alternative models of the origin of life have even more important implications. They endanger the very general understanding of life, of genetic determinism with all its extended analogies to culture. The implications might be paraphrased in the simple statement that life does not exist thanks to DNA molecules and the genetic code, DNA being merely one of a number of advantageous evolutionary leaps for information storage, including writing itself. The narratives of the holy scriptures are but a small portion of the overall narratological approach in evolutionary biology adopted by Markoš. In his understanding, life is a special kind of extended historiography:

> Evolution is a historical process analogical to that of cultural history: in short, neither past nor future development can be calculated, and the past can only be reconstructed, interpreted. All Darwinian biologists of whatever orientation are thus historiographers constructing a narrative based in scientifically proven resources. Comparative morphology, paleontology, or sequence comparison in molecular biology provide examples of scientific background for such narratives, but it is narratives that count, and their interpretation may substantially differ in various scientific schools. (Markoš and Švorcová 2019: 94)

Historiographical views on evolution are perfect examples of the interdisciplinary intersection between biology and the humanities; when methods from historiography are applied in evolutionary biology, the intersection becomes intertwined like the DNA molecule, at the moment when epigenetics retrospectively influences the humanities.

Cultural Epigenetics

As I already mentioned, epigenetics has been gaining a lot of attention from the humanities, mostly in history, cultural studies, cultural epistemology, and anthropology. Already before the arrival of epigenetics, there was a huge interest in genetics and in particular in the notion of Darwinian inheritance on the part of social science. I can mention the well-known term of "social Darwinism", which totally misinterprets Darwin's theory. The use of terminology from biology in the humanities and social sciences is very attractive, probably because it results in a seemingly more scientific approach which the humanities are accused of lacking. There also exists the opposite direction, merging the humanities with the concept of inheritance and the Modern Synthesis, which means adding human science notions to biological theories and not vice versa. I am referring here to the famous meme theory by R. Dawkins (1976).

There are already many existing different approaches of fusion between epigenetics and cultural phenomena. What they all share of course is the search for a happy continuity between culture and nature. This relation might be of very different characteristics and be supportive for many even dangerous ideologies and promotion of radical political or social theories (Dennis 1995). Paradoxically as it might seem, epigenetic inheritance applied to history studies can be, and by many is, associated with determinism (Jablonka 2016). It is only a paradox if we fail to consider the lawful part of epigenetics, discussed in the previous subchapter. In this perspective, as Jablonka notices, epigenetic inheritance, as any other kind of inheritance, is correlated with determinism because of the conviction that, once inherited, traits cannot be lost and remain stable, permanent and subjects to law. It is partially true that once a trait is inherited it can only hardly be lost during the lifetime of an individual.

The deterministic understanding of inheritance, so deeply anchored in the evolutionary theory of Darwin, or more accurately, in its later interpretation by the Modern Synthesis of Darwinism, and so scarcely challenged, could be an important topic for psychological or sociological studies. The popularity of social Darwinism, the everyday use of expressions like "it is in his genes", "I am genetically predisposed to…" exhibit the naive human inclination towards deterministic explanations. Whether it be religious faith or blind belief in scientific knowledge without further understanding of actual scientific method, determinism surely is psychologically convenient in the way that it deprives individuals of the weighty responsibility of free will, choice and decision making. Dawkin's theory fits this psychological explanation particularly well, but more about it later.

Epigenetics somehow disturbs the omnipotent power of genes and their irreversible and fatal influence on our bodies and personalities. This is also why it has many times been associated with the return to vitalism and Neo-Lamarkism. The accusation of vitalism is a weak defense of the dogma of the modern synthesis and its social and cultural implications.

But thanks to Peirce's message, we need no longer fear the unexplored lands of epigenetics. Scientific method in biology should rely on the triadic alliance of chance, law and habit. There, determinism plays an important role, yet it must be completed by the remaining two components Epigenetics positions biology somewhere in between lawful sciences (mathematics and physics) and human sciences, and understands biology as a metaphorical bridge between these sciences. Cultural epigenetics even more strikingly demonstrates this bridging and the interdisciplinary potential of biology. One may say, cultural epigenetics seeks patterns of change in society or culture analogous to patterns at the (epi)genetic level. The patterns can even be formalized or visualized diagrammatically, which will be the focus of Chapter 5. In what's left of this section I would like to more fully describe what precisely is meant by cultural epigenetics.

According to Eva Jablonka, cultural epigenetics demonstrates that "there is evidence suggesting that cultural practices lead to molecular epigenetic changes that in turn can contribute to the reconstruction of the system's dynamics" (Jablonka 2016: 46). By cultural practices Jablonka means for instance historical traumas, urban poverty, malnutrition; the consumption of unhealthy food, alcohol or other toxins; poor parenting; social inequality, bad schools; limited job opportunities; etc. Jablonka gives historical examples of individuals from:

> the Dutch Hunger Winter at the end of World War II, when the Nazi occupiers cut daily food rations to less than 700 kcal, 60 years later suffered from problems such as an increased risk of diabetes, obesity, schizophrenia and coronary disease… Compared with siblings of the same sex who were born during better times, they showed widespread differences in DNA methylation patterns, including at the insulin-like growth factor 2 (IGF2) locus. (Jablonka 2016: 50–51)

A similar epigenetic effect of the same historical trauma is mentioned by Trygve Tollefsbol:

> Human observational studies have focused on environmental stressors such as nutritional changes in the diet to study nongenetic inheritance of metabolic phenotypes. In one well-known cohort of pregnant mothers during the Dutch famine of 1945, babies born to mothers with low-caloric intake in their third trimester suffered from low birth weights and higher risks of mortality later in life. (Tollefsbol 2019: 199)

Jablonka additionally comments on many experimental studies both on humans and rodents where it has been observed that stressful or traumatic experiences such as "social defeat, a strong or enduring mental shock, physical and emotional abuse, or deprivation of early parental care can have deleterious long term, transgenerational effects that are mediated by molecular epigenetic mechanisms"(2016: 48)

Of course, the studies to which Jablonka and others refer in connection with cultural epigenetics illustrate how culture is inseparable from nature and how cultural evolution is influenced by an (epi)genetic pattern at the biological level, thus the interconnection between biology, culture, history, sociology and other sciences. Yet there is one additional point to be made. The epigenetic modifications mentioned by Jablonka are only secondarily cultural: they are all direct responses to environmental stressors such as lack of nutrition and so on, these having cultural causes, but which in relation to the organism's epigenetic response are irrelevant: given the same modification, the same response would occur whether the stressor be caused by a cultural event (war) or natural event (drastic climate change). Thus, the correlation between culture and nature in this case is not direct. I am not denying the interconnection between culture and nature in epigenetic modifications, but one should be careful when formulating conclusions. The field of study of cultural epigenetics is very young and we shall only see in the near future where it will be directed.

There is yet another possible understanding of cultural epigenetics. In the metaphorical sense we can understand it as intersection between nature and culture from an evolutionary perspective, comprehending human cultural history as a continuation of the evolution of species and having as the pattern the epigenetic forms, many cultural changes really being "written" in the methylated letters of the genetic alphabet. Besides the molecular patterns, the pattern might also be designed at the system level. Markoš and Švorcová talk about the constant fluctuation between plastic and static phases of systems in evolution, and these might be equally biological (as proposed by Czech biologist Jaroslav Flegr) and cultural:

> It is conspicuous that the curve is isomorphic for the evolution of cultures (Lotman 2009) and religions (Rappaport 2010), as well as life appearances (Flegr 2008, 2010); we have stressed the fact in previous publications (Markoš 2004, 2014, 2016, Švorcová 2016, Švorcová et al. 2018). It would appear that human cultural phenomena might represent but special cases of a general rule of the evolution of life, cases enabled by emergence of language. (Markoš and Švorcová 2019: 185)

As authors suggest, evolution of life can be assumed as general rule of which human cultural evolution is only one specific part. In this understanding, biological evolution is hyperonymic to cultural evolution, humans and culture being integral parts of living nature. Culture is thus not put in contradiction with nature. Like cultural evolution, biological evolution has not stopped; it may have stopped at the genetic level, but it continues at epigenetic level, which is both more rapid compared to genetic evolution and more flexible.

> The study of human epigenetic evolution could also contribute to our understanding of their genetic evolution. Genetic evolution during historical times was relatively rapid after the last ice age, 40,000 years ago, when it was driven by environmental fluctuations, population growth, and the colonization of new areas (Hawks et al., 2007; Wills, 2011).

However, it was probably much slower than epigenetic evolution: epigenetic variations are generated in an environmentally sensitive manner and at a much greater rate than genetic variation. Since genetic and epigenetic variations interact, the more rapid epigenetic variations could drive genetic evolution. (Jablonka 2016: 53)

"Human epigenetic evolution" mentioned by Jablonka is only a synonym for cultural evolution, which can be "marked" or "written" in our genome through epigenetic marks.

We believe that cultural and biological evolution can be put under a common denominator. What unites them is the simultaneous duration of the society and alternation of it. (Markoš and Švorcová 2019: 185)

Another relevant reference for cultural evolution is a paper by William Harms (1996) where the author presents an alternative to Dawkin's notion of cultural meme. It is not very clear what precisely is meant by memes, but in Dawkin's understanding memes are ideas, theories or religions which are passed to other individuals and generations. Both the spreading of genes and memes is explained through the term "replicator" by Dawkins. Genes and memes are shared through replication, and memes as the spreading of ideas are compared actual viral contamination.

What makes the idea of a "meme" initially so plausible is that cultural entities ordinarily seem to us to be "transmitted" through "imitation". We do have to make sense of the physical basis of cultural transmission. Doesn't this force us to assume cultural replicators just as the replicator theory does? Don't the very notions of "imitation" and "transmission" presuppose that there is some thing which is being copied or transmitted? The main purpose of this paper is to show how to understand cultural evolution without including any such mysterious things. (Harms 1996: 364–365)

As is clear from the quote, the notion of replicator presupposes that there is something to be replicated. And this *thing* is barely imaginable as some physical thing. In his paper, the notion "variable phenotype" is used instead. This alternative to the concept of meme has the advantage of not being dependent upon the notion of replicator and transmission. Harm represents cultural entities as a natural extension of the biological phenotype, as part of the biological regulatory mechanisms in higher animals. Additionally, Harm's modification to Dawkins' theory – or to the Modern Synthesis in general – is that he replaces the notion of "fitting" with the notion of "learning". Cultural entities do not fit the environment; rather, populations *learn about* their environment (Harms 1996: 370). The notion of learning is closer to Peirce's triadic evolution than to the Modern Synthesis, concretely reflecting the rules of Habit. The term "evolutionary knowledge" is also introduced.

> After all, if culture is the product of selection on humans and their ancestors, the reason culture exists in the first place is in order to increase our fitness. Put another way, perhaps cultural evolution just is us learning about our environment, as a group, in an evolutionary way, and that is what culture is supposed to do. (Harms 1996: 371)

In this quote it is elegantly illustrated how even the Darwinian notion of fitness can be used in the context of epigenetics or phenotypical variations without running towards deterministic conclusions. By learning about the environment, by creating cultural habits based on our interaction with the environment, we are learning and approaching scientific knowledge, and science in this way is also part of (cultural) evolution exactly as Peirce defines the term, as "knowledge gained by systematic observation, experiment, and reasoning ... the prosecution of truth as thus known, both in the abstract and as a historical development" (CD: 5397, accessed 25/2/2024).

Semeiotic Implications

The enormous implications of epigenetics in both the cultural and natural spheres of science are directly related to the notion of semeiosis. My aim has been to proceed from the scientific disciplines and arrive at the definition of semeiosis indirectly, as a logical epiphenomenon of scientific reasoning, exactly as it was for Peirce himself: Peirce only finalized his semeiotic doctrine after a lifelong study and contributions to many scientific disciplines (mathematics, chemistry, physics, logic and many others). The hope of this book is to demonstrate the strength and timeless importance of Peirce's ideas without recurring to clichés concerning Peirce's semeiosis doctrine. Semeiotic is defined by Kenneth L. Ketner as "the study of triadic sign relations (not fundamentally or merely the study of types of representamens)" (Ketner 2011a: 381). The triadic relations are originally logical relations, and thus by extension semeiotic:

> Later in his career, Peirce identified Logic with Semeiotic. This means in retrospect, one can revise his early phrase Logic of Science in terms of his later preferred terminology, to read Semeiotic of Science which one tends to shorten to just Semeiotic. He made it quite clear that Semeiotic was conducted only by scientific intelligences. (Ketner 2011a: 378–379)

There is a connection between logic, semeiotic and scientific reasoning in Peirce, where scientific reasoning should be supported by experimental results. Ketner even talks about the need for a "laboratory mindset" (2011a: 379) for semeiotic purposes. For this specific understanding of semeiotic, epigenetics appears to be a field where the veritable semeiotic method is used. As a derivative of genetics, since its very beginning epigenetic research has been related to experimental evidence and laboratory praxis. The discovery of the huge amount of regulatory genes was an example of a pure scientific method by abductive reasoning. A hypothesis, or rather a doubt was formulated by scientists about the uselessness of what was referred to as "junk DNA". It was improbable that such a great percentage of the DNA present in the nucleus of every cell of our bodies would be only a historical residue. Thanks to laboratory examination, evidence was found for the actual biological functions of the non-coding part of the DNA. The doubt was replaced by belief as it is described in Peirce's *The Fixation of Belief*. The relational character of molecular processes at the genetic level was discovered. What was formerly believed to be genetic causation limited to unidirectional dyadic relations was replaced by a pluri-relational dynamic interplay between genes, environment, behavior, individual organisms and lineages.

Relations go from the genetic script in both directions, horizontal and lateral on one hand and in both forwards and backwards directions on the other. By backwards direction we can understand the retrospective activation of previously silenced genes for example, or the reprogramming of epigenetic methylations, etc. These relations can be called semeiotic relations because they are based on an elementary triadic relation between chance, law and habit, that is, between a variable element (given by the unpredictable character of nature), a constant element (genetic determinism assured by the norm-code), and a historical-developmental element related to the notion of growth (the natural tendency to form habits and epigenetic interpretive interplay). The three elements can be accordingly substantiated in any triadic relation between a representamen (R), an object (O) and an interpretant (I), and this substantiation can be called a semeiosis. Epigenetic interplay between semeioses in unlimited semeiosis process has no beginning and no end. The starting point can be chosen randomly. We can start at the methylated mark of one of the genetic base "letters"; we can start at an environmental change; we can start at the birth of an organism. Regardless of where we start the description of semeiosis, it will always be a chain of relations we can call interpretive, coming from representamen R1 to object O1 through interpretant I1.

The difference between the Modern Synthesis model and EES models lies in that for the Modern Synthesis of Darwinism the semeiotic chain would stop at O1: a gene (R) is a representamen of a trait (O) bound together by the force of evolutionary habit (I). This habit can become frozen, almost like a Law, and it is a habit toward an ideal final interpretant. EES on the other hand does not stop at I1 but allows for I1 to become another object, O2 leading to another interpretant I2, and so on. For instance, a given gene might be translated into a peptide sequence leading to the expression of a trait, a particular wing color of a butterfly for example, due to the habitual hereditary process. The environmental conditions during the lifetime of the butterfly might change, the temperature might get significantly lower. When the new generation butterfly is born with the same genetic information as its parents, the gene is marked epigenetically due to the temperature stress. The expression of the same gene (but epigenetically modified) results in a different wing color or pattern. This semeiotic model is already significantly different from that of the Modern Synthesis. It no longer is a chain, since we must consider not only vertical hereditary relations (parents-offspring) but also horizontal relations of influence by external factors. Moreover, the interpretant I1 is not final because it is transformed into another object O2. The particular relation between I1 and O2 is complicated because of the immense complexity of the genetic information encoded in a single gene and the number of processes involved in the expression of the gene. It would be probably more accurate to describe every single step in protein synthesis (DNA-RNA transcription, splicing, translation into peptide chain, protein folding, chaperon work, etc.) as a singular semeiosis relation with individual characteristics, but it is not the aim of this book to provide a detailed semeiotic analysis of molecular processes. We can simplify for the sake of illustration only the relation between genotype and phenotype as a relation between representamen and object, where the interpretant is assured by the evolutionary habit. In this simplified model the relation between R and O is primarily indexical, given the genetic determinism and the understanding of the phenotype as an embodiment of material traces left after the previous generations.

But there is more than one particularity I would like to stress; indeed, attributing a purely indexical character to the relation between R and O in the (epi) genetic relations would be imprecise. Firstly, the relation between genotype and phenotype is partially iconic because of mediated resemblances, morphological and otherwise, of the embodied forms from the genetic script of the parents to the offspring. Secondly, this relation between genotype and phenotype is to a great extent symbolic because of the arbitrary and *digital* nature of the genetic script. We should keep in mind that notwithstanding the fact that the DNA molecule is a physically existing material molecule, the information it mediates and stores is purely digital, and the information becomes materialized only when the genes are expressed. Until that moment, information remains digital and encoded in a long sequence written in the genetic alphabet. Thirdly, the genotype-phenotype relation is symbolic in terms of thirdness and relation to the future, because referring to a future Object. This particularity concerns the possible existence *in futuro* of the actual object of the representamen, differently from the classical situation where object refers to its representamen retrospectively. But generally speaking, the most prominent characteristic of the semeiotic relation between genotype and phenotype is in terms of indexicality or indexical semeiosis.

Indexes are references to an object on the basis of a causal relation, almost like the algorithmic implication of the scheme *If A then B*, which can be illustrated as "If there is smoke, there must be (have been) fire", or "If there is a footprint there must be (have been) a foot". The verbal tenses are important to consider in these examples since indexes usually refer to the present or recent past, to objects close in time and space. In the case of epigenetics, the vicinity in time and space is more striking than it is in the case of the Modern Synthesis interpretation of genetic transfer, where the genetic information would be eventually expressed in the same manner regardless of what point of the time scale it is situated on. We can conclude that the epigenetic is in this sense more indexical than the classical genetic inheritance, which would be more symbolic because dependent only on the arbitrary and conventional genetic code.

When it comes to the semeiotic implications of cultural epigenetics in particular, we can say that also in this case we are dealing with primarily indexical semeiosis. Eva Jablonka predicts the use of epigenetic research in archeology in order to extract epigenetic information from ancient bones, providing information about nutrition and other important factors readable from the epigenetic script. This experimental use of epigenetic laboratory methods is likely to take place in future studies in human sciences such as archeology. From the semeiotic viewpoint it would be surely interesting to see the indexical pathways leading from (epi)genetic script to historical implications and interpretation. We can see an epigenetic mark as an index of the material traces of an object directly on an individual (the material object might be the contact with environmental stressors of the time the individual lived in: environmental temperature, diet, etc.), which consequently becomes another index for the archeologist-interpreter. Interestingly enough, this kind of indexical semeiosis loses the typical characteristic of a representamen being causally or physically related to its object in close vicinity in time and space (smoke persists only as long as, or a little after the fire is taking place). This kind of indexical semeiosis must necessarily be classified as indexical dicisigns (propositions), similarly to fossils in archeology: fossils are also indexes which have lost their feature of correlation to the present. Instead, they refer to the past, and in referring to the past they bear some typical features of iconic semeioses. This is a perfect illustration of the impossibility of clear distinction between various types of semeioses.

To summarize, gene expression sometimes depends on the environmental influence during the organism's development. Eco-evo-devo studies prove it, and this fact guarantees the semeiotic character of life. The concept of memory is also

important. Epigenetic memory is the temporary storage of particular information and it can be regarded as a minimal condition for human memory, which
can be extended to the historic memory of a whole nation or culture. This is the
way epigenetics becomes an interdisciplinary bridge between nature and culture.

> The humanities may point towards a general theory of evolution valid for all life. What
> human cultures have in common with the biosphere is the semiotic character of indi-
> viduals as well as communities, a character born of similar entities, which maintain the
> continuity of lineage. This allows, both individuals and communities, an interpretative
> approach to their history-rooted in memory and experience, yet also furthering their
> unique, individual, and creative approach to their genetic endowment, which is the cre-
> ation of novelty and the fact of evolution. (Markoš and Švorcová 2019: 188)

Epigenesis is triadic and relational, adding one more relate to the "frozen" dyadic
relation between the triplet alphabet of the genetic script and the phenotype[20].
Moreover, this relate is also determined by a kind of code – it is already a result of
habit formation in nature because it has a specific set of marks, epigenetic marks,
or epigenetic "signs", methylation marks for instance. Thus, even in the case of
epigenetics we can talk about a code, even if the code is not fully established
and as is not universal (it might be transmitted and might not, and if it is transmitted, it is potentially only to the next generation, and to a maximum of the
next five generations, but not further). More about the forming of code/langue/
structure will be said in the next chapters. Epigenetics is like the route from a
representamen to the final interpretant; very distant and never actually reaching
it. Epigenetic marks allow for deliberate changes without the need to remain in
the concrete state: if an environmental change occurs, epigenetic marks can be
demethylated or not transmitted. Epigenetic marks are real biochemical semeiotic relational units.

Lamarckism from Today's Perspective: Transgenerational Inheritance

Transgenerational inheritance is a special kind of epigenetic inheritance which
conserves acquired features for future generations (not only the next generation
but succeeding generations as well) even when the external stressor is not present. Transgenerational epigenetic inheritance can be explained in contrast with

---

20 The biochemical process is more complicated though: there is no unique interpre-
    tation by a unique "interpreter". Epigenetic marks are recognized by a different pro-
    teins than the genetic script interpreters. It is a complex set of multiple simultaneous
    interpretations.

generational epigenetic inheritance, where the acquired features are transmitted by epimutations on the genome from parents to offspring, for instance from mother to embryo, when some changes on the epigenome are induced by the diet of the mother during pregnancy. Transgenerational inheritance can persist even in generations 3 and 4. Sometimes the term epigenetic transgenerational inheritance is used in a narrow sense also to indicate only mother to child transition of epimutations (Tollefsbol 2019: 18).

The first reported study of transgenerational inheritance dealt with male fertility in rats, where impaired fertility and low sperm motility was transferred epigenetically to following generations, but only through male line (Tollefsbol 2019: 204). Probably transgenerational epigenetic inheritance is the only field of studies which can rightfully bear the name Neo-Lamarckism. Yet the term is already associated with a different branch of studies, mostly related to vitalistic pre-Darwinian views on evolution. But as was argued in the previous chapter, Lamarck's own ideas are quite distant from how he has been interpreted. The principal feature of "Lamarckism", the inheritance of acquired features, is now reborn under the field of transgenerational epigenetic inheritance.

Biochemical epigenetic marks are indexical signs of an intimate legacy with our parents. Again, the indexical nature is only a prevalent part of more complex semeiosis character of epigenetic signs, completed by the iconic and symbolic components. Iconicity is rather obvious when it comes to genetic script, given the complementarity of the bases in the process of transcription. The symbolic part is the most controversial, and even if very clearly stated by the definition of the genetic code, many scholars claim the absence or lack of symbolic representations in simple biological signs (for instance U. Eco 1999). In the understanding of evolution as a narrative pattern as proposed by Anton Markoš, the symbolic nature of semeiotic processes at the molecular level is undeniable. The genetic code was established by force of habit, which can be defined rather as law-like, dyadic, and thus an indexical mechanism regardless of the importance of its arbitrariness and conventionality. But when we add the epigenetic special marks to the extended understanding of the genetic script, its indexical causality does not hold. Epigenetic modifications are results of habit making and are what the genetic code was before it became a frozen, semi-automatized process. I argue that (in continuity with the Non Reduction Thesis, which will be presented in the next chapters) laws are derived from habits, anancasm is derived from agapasm, dyadic relations are derived from triadic ones.

Thus, in line with Peirce's evolutionary thinking, no such understanding of semeiosis is possible as a merely hierarchical scaffolding of symbols from indexes. Due to this reason, from a semeiotic viewpoint, the metabolism-first theories about the origin of life on Earth are more plausible compared to information-first ones: information storage and replication is already a dyadic, semi-automatic process which could have only derived from the previous habit-like behaviour of matter. Transgenerational inheritance works by indexical mechanisms: we are the footprints of our parents. It is important to realize that this observation is dramatically different from classical genetic inheritance, where the metaphors are used comparing phenotypes with blueprints of genes. In genetic inheritance, no matter what the organism experiences, s/he transfers the same genetic information that has already been transmitted for thousands of years. The organism is only a container of replicators. It has no active role in interpreting the genetic script. This kind of indexical inheritance is indeed dyadic whereas in the case of transgenerational inheritance organisms interpret the genetic script and pass it to the next generation, but with their own modifications, with their own life experiences, thus these are indexes already closer to symbolic thirdness – these indexes are genuine *semeioses* in that they are triadic, containing the interpretant part responsible for the thirdness.

Thirdness is also related to the notion of memory. Symbolic semeioses are characterized as making reference to the future, in contrast to iconic semeioses that refer to the past and indexical semeioses that refer to the present. Symbols have the capacity, thanks to their arbitrariness and correlated conventionality, to refer to future objects without actual physical or causal bonds; but in order to function, memory is required for the semeioses users. They must have the capacity of storing the conventional relation of the representamen with the object. In living organisms, memory is also present. We can talk about "epigenetic memory", which is necessary because epigenetic semeioses are not part of the frozen genetic information stored in the DNA molecules; the storage must be provided otherwise. Trygve Tollefsbol states that epigenetic inheritance is in fact this memory, needed to preserve epigenetic marks for future generations:

> Epigenetic inheritance is a form of genetic memory storing cues from environmental stressors in previous generations in the DNA of future generations. Transgenerational inheritance occurs without the persistence of the environmental stressor, and the most widely referenced hypothesis indicates the germ line for mediating inherited marks. The postulated mechanism suggests that environmental factors promote changes in epigenetic patterns, or epimutations, that become imprinted like in the germ cells and escape DNA methylation erasure during epigenetic reprogramming. (Tollefsbol 2019: 203)

Because epigenetic reprogramming can erase the epigenetic patterns, the epigenetic marks can be *forgotten* in case of epigenetic inheritance. Yet in the case of *transgenerational inheritance* the inherited traits can escape the erasure, which guarantees longer epigenetic memory, which endures even when the indexical environmental stressors are gone. Transgenerational epigenetic inheritance is consequently the real guarantor of the triadic relational aspect of the molecular processes responsible for gene expression. This kind of memory replaces the need for the frozen final interpretant of the genetic script and allows for triadic interpretive processes at the lower level of living beings. The difference between genetic inheritance and transgenerational epigenetic inheritance is, in semeiotic terms, the same as the degenerate dyadic versus genuine triadic semeiosis. Epigenetics is where the molecular interpretive semeiotic processes take place thanks to the action of habit. But as in the case of all other kinds of semeiosis, the triadic is impossible without the dyadic, because every triad is decomposable to dyad. In the same sense, epigenetic semeiosis is impossible without genetic semeiosis:

> It would be misleading to even compare these two inheritance systems in terms of importance: genes are the major source of inherited information, but the genetic and epigenetic level should not be considered in separation, although one is more plastic than the other. These two inheritance systems are essentially interconnected and work always in concert, and although we are accustomed to it, there is nothing inherently correct about prioritizing genes and seeing epigenetic processes as somehow less important. (Švorcová and Kleisner 2018: 233)

Genetic and epigenetic semeioses are thus not to be understood in contradiction but rather as in complementary relations. According with Peirce's Law of Mind (1892) (Bisanz 2009: 83–85), there is continuity between semeiosis relations in the whole universe and semeioses are only complete and real when related to other semeioses. Anancasm is degenerate agapasm, in the same way genetic script as immutable set of strings is also a sort of frozen habit, a rule resulting from habit taking becoming regularized or ossified. This rule can be ultimately modified is external factors lead to such modification, with aid of a different kind of coding, non-frozen coding or epigenetic marking. Signs are relations and semeiosis extends relations to other relations. Biological inheritance can be understood as such relations between the genetic script, the organism, the environmental and other external conditions. Cultural evolution is in continuity with biological evolution. Culture can influence our (epi)genome and vice versa, our biological limitations influence culture. Thanks to research in epigenetics we arrived to the point where the difference between nature and culture is erased, which is one important argument in favor of the hypothesis by Ketner against the current clear cut in academia between natural science and humanities/social science, in author's words, against scientism (Ketner 1999). Inheritance regards not only material physical features like eye or skin color, but also non-material realities, such as fears, phobias or ability to sing or other particular talent. Non-material realities are relations and are part of scientific inquiry. The progress of current science is leading us towards the inseparability between science and humanities, so that semeiotic is also understood as interdisciplinary inquiry about relations, as Peirce discovered.

Chapter 4 Manifold Proteins: Diagrammatic Models for Protein Folding

Proteins are Folded Strings

Peirce was fascinated by biological phenomena, mostly by the protoplasm and neural system, and he surely would be fascinated by the discoveries of the last decades, starting with the deciphering of the genetic code and continuing in the direction of the discoveries of the proteid structures, of which he himself predicted the importance and complexity:

> The chemical complexity of the protoplasm molecule must be amazing. A proteid is only one of its constituents, and doubtless very much simpler. Yet chemists do not attempt to infer from their analyses the ultimate atomic constitution of any of the proteids, the number of atoms entering into them being so great as almost to nullify the law of multiple proportions. (EP: 267)

Proteins have been fascinating for scholars since before the deciphering of the genetic code and the related knowledge about their physico-chemical properties and the process of their biosynthesis. What has always been captivating about proteins is their structure, in Peirce's words their "atomic constitution". Not only biologists, but also philosophers are tempted to study protein structures and the whole process of their production. From a semeiotic viewpoint, proteins can be seen as more basic protosemeiotic agents (Sharov 2010, 2015). Proteins are the chief protagonists of this chapter as perfect examples of minimal semeiotic biological structures whose functions are irreducible to explanations by dyadic relations. This is evident due to the very structure of folded strings (peptide chains). Let me start with the following question: what are proteins and why are they interesting for theoretical inquiry? I will briefly explain the basic principles of protein structures and protein synthesis in this and the following sections.

Proteins, the smallest functional units of our bodies, are complex organic macromolecules (see Fig. 2). They probably existed at the very beginning of life on Earth[21]. Many discussions about the origins of life or about the origin of the

---

21   The origin of life on Earth is unknown, yet the most plausible hypotheses, so far, are of two basic types: it was nucleic acids first (the code) or it was proteins first (what the code codes for). The supporters of the protein hypothesis are not rare, e.g. (Andras and András 2005). Besides code and proteins, there is also a third option, the so called "RNA world hypothesis" (Robertson and Joyce 2010). For more detailed excursion

genetic code are intrinsic to the discussion about the origin of proteins (Crick 1968). In fact, the very reason for the existence of the genetic code rests on the need for the synthesis of proteins. Our cells use the genetic code to preserve genetic texts (scriptures) with information about which proteins to produce and how to produce them. The miracle of life is hidden in protein synthesis.

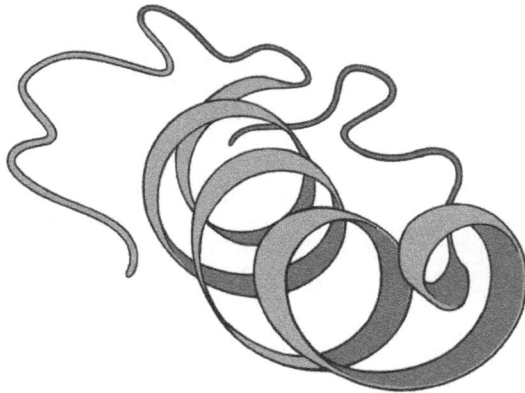

Figure 2: Protein structure.

Proteins are the elementary building blocks of life; they constitute cells and fulfill of all metabolic processes, they function as antibodies to protect organisms against diseases, they regulate gene expression etc. From a chemical viewpoint, proteins are but folded strings of hundreds of amino acids. An unfolded string is just a long chain of molecules, yet the moment in which it folds and deflects into a compact bunch, surprisingly a three dimensional and functional structure, which guides all organic function in our bodies, appears. Proteins have been an interesting topic for philosophical essays since the very beginning of modern molecular biology and genetics (Jacob 1970; Monod 1972; Deleuze and Guattari 1987): the process of change from the linearity of the original peptide chain to the dimensionality of the final protein has been especially stimulating for fruitful philosophical discussions.

---

into different hypotheses about the origin of life on Earth see Markoš and Švorcová (2019).

In the last few decades, the quest for protein folding has been in the center of scientific inquiry more than ever before in the history of biology. Thanks to progress in informatics and newly created fields such as proteomics or bioinformatics, scientists have strong tools to quantitatively analyse a huge amount of data and to try to predict protein structure exclusively from the sequence of amino acids within a peptide chain. For these purposes, linear transcriptions of peptide strings stored in protein banks (some of them available for free) are used. Big data analysis with the newest quantitative methods, such as n-gram comparison (Faltýnek, Matlach and Lacková 2019) and the bag of words model (Owsianková et al. 2020) provides scientists a new form of experimental approach to organic strings, when transcribed, based on understanding strings as virtual texts and therefore analyzable with big-data methods. However, we do not know by what rules or laws strings become structures, that is, in the case of proteins, according to what rules amino acid chains turn into three-dimensional protein structures. Protein folding, a process of getting to the final protein structure from the original peptide chain, is still a very obscure process, and the procedures governing protein folding remain to be discovered. Perfect knowledge of the chemical and physical properties of the peptide chain with the potential to fold does not help in understanding the question of why proteins fold in that way and not another. Chemistry and physics play a role, yet don't seem to be the only answer. This requires an explanation of another kind: "protein code", "protein grammar" and "protein syntax" are terms which occur: "a code" is being sought rather than a purely chemical explanation. This situation is very similar to the deciphering of the genetic code, as it is believed that protein folding is a mechanism that was obtained by natural selection, meaning that it was achieved in a way similar to the genetic code, by evolutionary convention rather than chemical or physical necessity. The relation between string and structure is a relation between the dyadic and the triadic. I will address the question of this relation in the following pages.

From an evolutionary viewpoint, protein structures resulted as a response to environmental stimuli and were conserved in the following generations, becoming habits. The possibility that protein structures somehow preexist in nature in the shape of Platonic forms is not maintained by today's scientific mainstream; it is believed that protein structures are rather results of various evolutionary factors such as fitness and thermodynamics:

> An understanding of the interplay of protein structure with both sequence evolution and functional/phenotypic evolution is necessary [...] To rigorously evaluate the possibility of a fold transition one would have to determine the viability of a series of mutations that connect the two folds. Both thermodynamics and kinetics of folding must be

taken into account, as well as fitness effects due to function, all within in a context of population genetics. (Siltberg-Liberles, Grahnen, and Liberles 2011: 749–752)

Yet some scholars do believe in Platonic forms as preexisting in nature and in the evolution of protein structures (Denton 2002; Marshall and Legge 2003), raising the metaphysical problem of the persistence of the Platonic explanations in biology already discussed in previous chapters. In protein studies, the Platonic view on biological macromolecules is particularly tempting. It is worthwhile to recall the agapastic evolution by Peirce, who actually reckoned Platonic forms in evolution:

> The evolutionary process is, therefore, not a mere evolution of the existing universe, but rather a process by which the very Platonic forms themselves have become or are becoming developed. (CP 6.194)

The present perfect and present continuous verb tense is an interesting choice by Peirce, which effects the understanding of "Platonic forms" of life as constantly being under development, thus not predetermined and not working as mere blueprints of organisms, in our case one might say, blueprints of proteins. Despite the fact that he uses the term Platonic, his argumentation is distant from the classical understanding of Platonic forms. There is a gradual passing from habits to frozen code, from the physical world to abstract forms: chance in the case of evolution is different from chance in the throw of the dice. While the former is arbitrary in the sense of spontaneous choice between numerous options, the latter is arbitrary in the sense of pure coincidence. I already mentioned the tension between Platonic forms in biology and Darwinian evolution in Chapter 2. Seemingly outdated Platonic explanations in biology have emerged many times after Darwin, but the present search for Platonic forms is even more striking with the recent arrival of proteomics and the exploration of the limited number of protein structures. Markoš and Švorcová (2019: 16) connect the current research in proteomics with pre-Darwinian Platonic movements in biology when they speak about the search for "laws of forms". The authors develop a pleasant argument contra the Platonic view on proteins:

> If the environment produces random polypeptide strings packed into limited varieties of shapes; and
>
> - If there exists an environmental scaffold that recognizes not sequences but shapes of such polymers;
> - Then higher-order complexes should assemble repeatedly, according to the properties of the specific scaffold (be it, e.g., a mineral, membrane, prosthetic group, or RNA). What holds for polypeptides can be true for other polymers such as nucleic acids, lipids, or polysaccharides. As the environment can supply an enormous heterogeneity

of such matrices (see the examples of clays, minerals, and rocks in this chapter), a parallel testing of different assemblages might have proceeded – and selected all this organic manifold – without a need to store information in a medium different than structures themselves. (Markoš and Švorcová 2019: 16–17)

Thus, if there were pre-existing ideal proteic forms, there would be no need of the stored genetic script, in that case the evolutionary emergence of the medium for the storing of information would be unexplained and unnecessary. The limited number of existing proteic structures is an experimental fact; but the Platonic explanation is not the only possible one. As with any other organic forms, also proteic forms can be explained as results of natural habits rather than laws.

One important observation should be made before I go forward in my argumentation. I pointed out that in the case of proteins, it is the very process of folding that provides their functionality: in the form of a string, peptide chains do not warrant biological functions. Therefore, the folding, which is in fact a binding between singular amino acids of the peptide chain, is in the case of proteins crucial from the functional or semantic standpoint (see Fig. 3). This is how binding between parts of a peptide chain differs from other biological binding of strings, for instance the spatial arrangements of DNA. In the DNA molecule, two strands of the molecule twist around the helical axis to create the famous twisted double helix structure. Many further possible ways of compacting the string exist in nature. All possible spatial DNA structures are studied by the field of DNA topology, such as helices, loops or junctions. In recent years, the structures of supercoils of DNA have attracted a lot of attention from scholars; supercoils are the additional twistings of already twisted double helix. The difference between DNA topology and protein topology, simplistically speaking, resides in that for DNA the spatial arrangements help with economization of space, while in the case of proteins the spatial structures correspond to the biological function[22].

22 It has to be remarked that some DNA loops do have a biological function, most often a regulatory function.

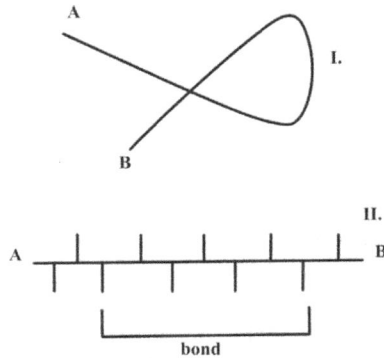

Figure 3: Biological function as folding. Part I corresponds logically to part II. Whether in the form of a string or in a loop-like structure, the information contained in the DNA remains the same. Binding between amino acids, on the other hand, corresponds to the creation of a biological function.

Protein Synthesis, Basic Principles

I outline here some basic biological principles necessary for understanding protein synthesis, for the sake of clarity. It is important to understand the basics of protein synthesis because protein synthesis is actually the very way of gene expression: a gene is expressed when its coding in the DNA is used as instructions about how to build a protein, thus, the actual protein corresponds to the gene encoded in the DNA. The DNA molecule is formed of four nucleobases (eventually five, because thymine is altered by uracil): adenine (A), cytosine (C), guanine (G), thymine (T), and uracil (U). In the process of protein synthesis, the DNA sequence is transcribed into the RNA sequence, which is in its turn translated into amino acids. The two aforementioned processes are accordingly referred to as transcription and translation.

Transcription

The process of transcription occurs thanks to a specific enzyme, RNA polymerase. RNA polymerase "opens" the double strand of the double helix of the template DNA so that it acts upon only one of the strands and is "read" base after base, pairing every time the complementary bases, with the difference that the base uracil appears instead of thymine. Complementariness is expressed in the pairing of purines (adenine and guanine) with pyrimidines (cytosine, thymine and uracil). Purines are complementary only with pyrimidines and vice versa, as pairing with only purines or only pyrimidines would be energetically unfavorable. The pairing between purines and pyrimidines might be called a "transcription code". The pairing takes place thanks to hydrophobic and hydrophilic forces between the complementary bases, creating hydrogen bonds, thus "purely physical" phenomena are going on when DNA is transcribed into RNA sequence. The transcription results in the RNA string which is a concatenated chain of paired bases. This RNA is single-stranded and we also call it messenger RNA or mRNA. It is called accurately messenger, because all the information encoded in the double-helix DNA is transcribed here and mRNA further serves as the actual messenger, a storage of the instructions for the protein synthesis, thus gene expression in the actual protein. This description is simplified for the sake of clarity and purposes of this book. But it should be at least mentioned that the whole process is much more complicated, for instance after the transcription the RNA string (primary transcript) is *spliced*, that is, irrelevant or unimportant parts are cut off, and as a result only the coding part of the RNA transcript remains. The parts of the transcript that are cut off are called introns while the remaining parts are called exons. As a consequence, many functional variants can be "cut" from one single gene (Reed et al. 2010).

Translation

Translation is the second fundamental step in protein synthesis. It is accurately named translation because the RNA transcript from the first step is *translated* into amino acids. These bind together one after another immediately after being translated, creating another string, a chain of amino acids also called the peptide chain. Analogically to the transcription process, where the RNA polymerase is needed, in the case of translation the process takes place thanks to another macromolecule, the ribosome. The ribosome provides the structure where the whole process of translation happens. In the case of translation there is a pairing between elementary units, between bases from the RNA string on the one hand and amino acids on the other hand. Watson and Crick (1953) demonstrated that this pairing is not a matter of one-to-one correspondence, but triplets of nucleic bases pairing with single amino acids. The pairing between triplets of nucleobases and amino acids is what is referred to as genetic code. What is so different between transcription code and translation code is that during transcription only hydrogen bonds assure the stability of the code, therefore, the question is whether it should be considered a semeiotic code at all. Unlike transcription, during translation the pairing does not depend upon physical forces, or, to put it more correctly, does not depend *only* upon physical forces: there is no direct interaction between base triplets and amino acids, they never touch one another. Instead, a "mediator" is needed, the molecule of tRNA carries out the function of a medium between the two elements of the code, having binding sites for both triplets and amino acids. At this point, it is convenient to comment shortly on the notion of the word "code" used in genetics. The notion of "code" can be understood in many ways. Some understand this word in the dyadic terms of a causal correspondence, or algorithm, and apply it mostly to digital or computer codes. A semeiotic specification of the term code from the theory of Eco is helpful here. He distinguishes codes proper from s-codes, where the latter correspond to the algorithmic and dyadic matching between two entities, and the former correspond to a triadic bringing together of two distinct worlds, of two separate s-codes by way of mediation. In the case of the genetic code the mediation is guaranteed by the tRNA molecule: transfer DNA, literally meaning that it transfers one part of the code to the other, creates a link between them but does not directly participate in the final matching; the role is only mediatory.

In these brief paragraphs I have summarized the process of protein biosynthesis in the most simple way. As the reader may have realized, it is very difficult, if not impossible, to describe the phenomena at the molecular level of life without a certain anthropomorphized lexicon or without expressing some level of agency. Due to the explanatory motivation, but not only, it is very common also in scientific jargon to rely upon such vocabulary. Many scholars have tried to escape from the trap of anthropomorphism in biology – the problem may not be in anthropomorphism per se, but rather in the goal-directedness that is associated with anthropomorphism. Italian biologist Giorgio Prodi turns the problem the other way around, putting in the center of the discussion lower organisms instead of humans, so that anthropocentrism becomes viewed from the opposite direction. According to Prodi, the statement that a cell or a ribosome *reads* is not to mean that it simulates the human act of reading, but rather that human reading has origins in molecular interpretative processes (Prodi 1989: 25–26).

I believe that Peircean evolutionary agapastic philosophy offers a feasible alternative to anthropomorphic definitions in molecular biology. The triadic logical system is a strong instrument for resolving the abiding problems of describing biological phenomena. It is possible to express the nature of life as non-dyadic relations between cause and effect, and at the same time remain aware of vitalistic and other non-scientific explanations. If the triad or teridentity relation is a genuine elementary logical relation, then it might be situated at the heart of scientific explanations in biology. It will be shown that the best way of formalizing teridentity is with a specific kind of graphical logical notation, which is presented below. Before I move to the formalization of protein synthesis, it is necessary to extend the short introduction to molecular processes by presentation of amino acids and protein folding.

Protein folding

Even though approximately 500[23] different types of amino acids occur in nature, only twenty of them were selected by natural selection to play a role in protein synthesis. After the translation process (illustrated in Fig. 4), the amino acids bind together in order to form a chain, a sequence called a peptide chain. The chain of amino acids holds together thanks to peptide bonds between every neighboring amino acid in the translated chain. The amino acid chain provides the material for the protein, which is consequently built up from the amino acid chain simply by its folding. Not all of the twenty amino acids are necessary to construct a protein; it was proven experimentally with artificially synthesized proteins that a small structure may be built up from a peptide chain composed of combinations of only three types of amino acids (Berezovsky, Guarnera, and Zheng 2017).

In agreement with traditional views on protein folding, the folding process has three steps. Firstly, the primary structure (the peptide chain) is folded to form secondary structures[24]. Secondary structures are consequently bound together to form the tertiary structure. Afterwards, it is possible to obtain the quaternary structure by combining two or more tertiary structures together.

Now let's have a further look at the amino acids. The amino acid is composed of

- a central carbon atom,
- a hydrogen atom,
- an amino group – consisting of a nitrogen atom and two hydrogen atoms,
- a carboxyl group – consisting of a carbon atom, two oxygen atoms, and one hydrogen atom,
- an R-group or side chain – consisting of varying atoms, also called residue.

---

23   See Cole and Kramer (2016). It should be noted however that this number is only approximate, and in fact there are some specific amino acids present in mitochondria (King 2007).

24   The importance of secondary structures in protein folding has been questioned by Berezovsky and Trifonov (Trifonov et al. 2001; Berezovsky, Guarnera, and Zheng 2017) who propose to consider the "closed loops" as having more importance in the very process of folding, secondary structures only playing a role in the final detailing of the protein.

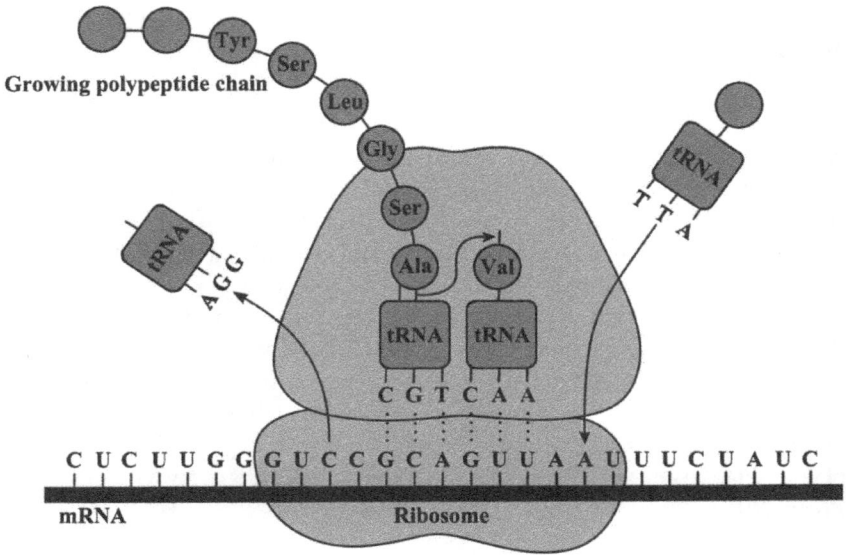

Figure 4: Translation. Peptide chain represented traditionally as a necklace, amino acids being single beads.

Amino acid residues are important because they are the special part which gives the amino acid its uniqueness. Amino acids bond together to form a peptide chain.

Usually a peptide chain is at least around 250 amino acids long. When amino acids are lined up to form a protein, they arrange themselves to form secondary structures. Two basic secondary structure types are known: alpha-helix, which is a coiled shape, and beta sheet, which is a zig-zag shape (see Fig. 5). While the folding of secondary structures is based on binding between close amino acids (hydrogen bonds are created), throughout the folding of the tertiary structure, amino acids bind the distant pieces of a peptide chain together (disulfide bonds or other strong bonds are created).

I have very briefly introduced the process of protein folding, but I did not mention why proteins are so important and what they are good for. Proteins are working components in our bodies and they have many functions in metabolism. For instance, they are the bricks that build bones and cells, they are transportation machines, DNA readers and interpreters, etc. All these various functions are due to the differences in proteins' shapes (structures). Protein hemoglobin, for

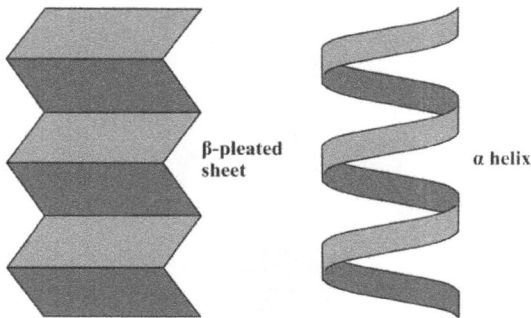

Figure 5: Protein secondary structures.

example, has a specific shape that fits to the oxygen molecule, as a key fits to a hole, and thus can function as an oxygen transporter. The metaphor of key and a hole is of course not exact. The fitting is not always perfect, as proteins "touch" each other or their target molecules and try to find the fit. Still, the crucial characteristic of the protein's function is therefore its shape, in particular, the way how the concrete part of its shape (the one that interacts with other molecules) fits to the target molecule's shape. In the case of hemoglobin, it is the binding or matching between the binding site of hemoglobin (the region of the protein entering into interaction with the target molecule) and the shape of the oxygen molecule. In biochemistry the interacting molecule is called the ligand. The complex of protein and ligand creates a biologically functional unit.

> Briefly, thanks to its specific binding center, a protein is able to recognize out of the background of thousands of different molecules, its *ligand* (a small molecule, a specific part on other protein(s), a photon of a given ligand wavelength, mechanical inputs, etc.). (Markoš and Švorcová 2019: 27)

Another example might be represented by enzymes. Enzymes are special proteins whose function is to speed up chemical reactions; they are biocatalysts. The catalysis works thanks to enzyme's active site, the part of the enzyme that binds with its target molecule (substrate) by shape fitting (as illustrated in Fig. 6). The shape of the enzyme's active site is the result of protein folding to that particular shape.

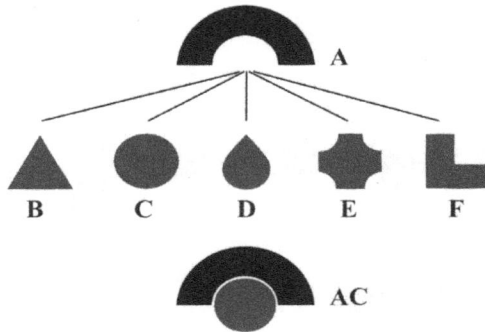

Figure 6: Enzyme binding to its substrate. Illustration inspired by Cimatti (2000, 52).

The unsolved problem in current research in protein studies is the process of transition from the unfolded sequence of the peptide chain into a unique three dimensional configuration. As a matter of fact, one single amino acid sequence may form different structures in different proteins, but also vice versa, the same structure might be encoded by different strings. Consequently, there is no such rule saying that every specific sequence corresponds to a specific shape. Additionally, the final shape also depends on many external factors. The following question unavoidably arises: why does a sequence fold in one way in one situation and in another in another situation? Or in other words: what is the relationship between sequence and structure? The question about the relationship between sequence and structure warrants research into thermodynamic explanations. A possible thermodynamic explanation would be that all possible constructions are tested to find the most energetically favorable one. But the amount of all possible constructions of one peptide chain is so high that it remains a mystery to understand how evolution came up with stabilized protein structures. It was calculated by Cyrus Levinthal (Levinthal 1969) that a random search for protein structure would take up to 5×10^{34} seconds, or 1.6×10^{27} years. The enormous difference between the calculated and actual folding times is called Levinthal's paradox.

Diagrams

The difficulties that arise from the linear understanding of the relation between the sequence and the structure of proteins, may be resolved by a diagrammatic or topological approach towards this relation. Topology is originally a mathematical concept, but can also be used as a theoretical and explanatory position towards any scientific discipline. Topology is a concept present in the philosophy of science already for several decades and it is an alternative instrument in scientific explanations permitting non-linear representations, that is, representations which are at least two dimensional, and thus may avoid the trappings of temporal and sequential representation. In other words, when not sequential, the order between the two elements entering the relation might be reversed, or simply not determined, or simply inexistent, which would mean that the two elements coexist without a specific order or hierarchy between them. Topological explanations have been very popular in the philosophy of science in the last few decades (Lacková and Zámečník 2020), but still we can say that they are not new. In some disciplines topological thinking existed long before the term of "topological explanations" came into use. For instance, in linguistics the structuralist era can be understood as topology applied to linguistic systems (Lacková 2021).

In some fields of biology, topological explanations have also already appeared. For instance, within the field of evolutionary biology, the relation between genotype and phenotype has been approached from a topological standpoint. By genotype we understand the whole inherited genetic information of an organism, while by phenotype we understand its actual manifestation with observable characteristics in the form of an organism. The classical dyadic and linear explanations comprehend phenotypes as simple blueprints of genotypes, meaning that they presuppose a dyadic and linear unequivocal correspondence between genotype and phenotype. As an alternative, genotype-phenotype "maps" (G → P maps) were proposed to diagrammatically track the whole biological passage from digital script to a three-dimensional bodily structure in a diagrammatic way (Huneman 2010).

As Massimo Pigliucci noted:

> the undeniable progress we have made in understanding G → P maps, both empirically and theoretically, is such that one should hope that evolutionary biology has reached the point of forever being past simplistic ideas like genetic programmes and blueprints, embracing instead a more nuanced understanding of the complexity and variety of life. (Pigliucci 2010: 564)

Elsewhere Pigliucci explains the importance of the concept of genotype → phenotype mapping and the need to abandon the gene-centralized model:

> Genomics and what I refer to as "postgenomics" (proteomics, metabolomics, etc.) started out squarely within the conceptual framework of the rather gene-centric MS, with the view that once we "decode" the genome of an organism we somehow gain a universal key to understanding its biology. The reality of organismal complexity has

shattered such simplistic visions [...] The complexity of the genotype → phenotype map cannot be understood only by bottom-up approaches such as those that focus on gene networks and regulatory evolution, however. (Pigliucci 2009: 223)

Topology is central for Peirce's writings. Mathematical notions of continuum, diagrams and topology enter into Peirce's accounts on mathematics, but are also essential for the current investigation in biology. Diagrams and topology are already widely used in proteomics, while continuum is too complicated to be comprehended by classical biological paradigms. The next chapter is dedicated to the possibility of application of the continuum (in mathematical and metaphysical understanding of the term) to biology, in addition to the biological applications of topology and diagrams presented in this chapter. Diagrams are for Peirce understood as part of mathematics (Scott 2006). Mathematical reasoning in terms of diagrams and topology is the precondition for Peirce's later semeiotic and his doctrine of semeiosis. According to Ketner and Putnam, mathematical reasoning is directly related to Peirce's semeiotic and pragmatic maxim:

> Because it (mathematics) explored the consequences of pure hypotheses by experimenting upon representative diagrams, mathematics was the inspirational source of the pragmatic maxim, the jewel of the methodological part of semeiotic, and the distinctive feature of Peirce's thought. As he often stated, the pragmatic maxim is little more than a summarizing statement of the procedure of experimental design in a laboratory – deduce the observable consequences of the hypothesis. And for Peirce the simplest and most basic laboratory was the kind of experimenting upon diagrams one finds in mathematics. (He understood the word diagram broadly, encompassing visual, tactile, or audio entities used to model a set of relations under study.) (Ketner and Putnam 1992: 2–3)

Diagrams are thus understood as mathematical notions. But it is important to define what was Peirce's vision of mathematics. Mathematics for Peirce was not necessarily about quantities, as it was still common to understand mathematics in the times of Peirce (Scott 2009: 102). Mathematics was for him more about spatial relations, geometry and topology. Peirce introduced a rather relational definition of mathematics, compatible with what we call topology. Peirce was certainly interested in the mathematical dimension of topology. Topology as a mathematical doctrine did not yet exist in Peirce's times, but there were already notions like Listing numbers (the name itself proposed by Peirce) to denote spatial relations in geometry. Peirce used the term "topics" and one of the examples of Peirce's interest in topology can be found in *The Basis of Pragmaticism*, p. 263, or in his correspondence with Lady Welby (Hardwick, ed. 2001). Peirce defines "topics" as the order of connections of space (Welby-Peirce correspondence, 48).

The four aspects of space are the Listing Numbers and are given special terminology by Peirce: Chorisy, Cyclosy, Periphraxy and Apeiry. Apart from Listing numbers, seemingly the most crucial topological question for Peirce was the so-called "map-coloring problem" about the definition of the minimal number of colors to delimit boundaries of singularities within a given space (see comments on the problem in many papers by Peirce published in NEM, all volumes; for a general account of Peirce's topology see Havenel 2010). I will not go into the mathematical details about "topics" in Peirce even though its importance in the field of mathematics is undeniable. We can say that by simplification, a graph is defined by Peirce in topological terms, meaning that it cannot collapse by any motion as far as the relations between parts of the graph remain:

> Let a particle have space within the colored parts of the graph to expand in M-A dimensions, so as to form <u>unlimited</u> (but not generally infinite) M-dimensional continuum (like a ring or a closed surface), which, once formed, can by no motion within those closed parts of the graph without rupture collapse. (MS 482: 1)

Topology is a part of geometry dealing with relational aspects of points of a space regardless of what the actual points are or represent, that is why it is a universally applicable doctrine. Yet it is important to stress the mathematical basis of the topology later applied to many scientific fields. Indirectly, topology is present in all Peirce's writings. I will argue that topological thinking is a characteristic feature of Peirce's whole body of work, concretely supported by the highly developed visual manifestation in the form of diagrams.

Besides his explicit mentions of "topics" in mathematics, Peirce applied topological thinking to his diagrammatical logic[25]. It is widely recognized (Stjernfelt 2007: 82) that Peirce's works on diagrams are partly derived from Kant's schemata (from *Critique of Pure Reason*). It can be said that for Kant, the notion of schema is a matter of relation between two very different kinds of things, concretely concepts (or categories) and sensations. Concepts have no direct connection to sensations and yet they are somehow related: sensations are only cognizable thanks to concepts and the other way round, categories are only possible thanks to sensations.

But how is it possible that non-empirical concepts are related to empirical sensations if the two differ in the very nature? Kant uses the notion of schema to resolve the problem of relatedness of a-priori concepts with empirical sensations. The element that the two totally different things share is the schema, something that belongs to both. In the words of Kant:

---

25   I am very thankful to my friend Diego Gabriel Krivochen, a brilliant mathematician, for the clarifications and consultation on topology.

> the categories, without schemata are merely functions of the understanding for the production of conceptions, but do not represent any object. This significance they derive from sensibility, which at the same time realizes the understanding and restricts it. (Kant 1781: 80)

A schema is something that categories share with sensations, something in which the two resemble each other, but what a schema really is for Kant, is a very abstract concept. It might be surprising at first glance that a schema is not an image; it has no visual existence, yet is a predisposition to create visual images. It is a "schematism of pure understanding" (Kant 1781: 77). The schema differs from the image in that the latter precedes the former.

> The schema is, in itself, always a mere product of the imagination. But, as the synthesis of imagination has for its aim no single intuition, but merely unity in the determination of sensibility, the schema is clearly distinguishable from the image. (Kant 1781: 77)

The relation between Kantian schema and Peirce is summarized by Marc Champagne (2018). To explain better the difference between image and schema, Kant proposes the example of a triangle, where the cognitive idea of a triangle is related to its schema rather than to its visual image. This is due to the generality of the idea of a triangle: while an image must necessarily always be designed in a determined way, whether it be an acute-angled triangle or a right-angled triangle, this is not the case of a schema, which in its part is not limited by a necessity of concrete representation because it only exists in thought:

> No image could ever be adequate to our conception of a triangle in general. For the generalness of the conception it never could attain to, as this includes under itself all triangles, whether right-angled, acute-angled, etc., whilst the image would always be limited to a single part of this sphere. The schema of the triangle can exist nowhere else than in thought, and it indicates a rule of the synthesis of the imagination in regard to pure figures in space. (Kant 1781: 78)

Now, as will be explained below, Peirce's diagrams in terms of his diagrammatic logic are visual diagrams used to express logical relations[26]. This does not however mean that they are images. The visual representation of diagrams is only a substantiation of the conceptual diagram responsible for determining logical relations. Diagrams are general significations, their general character is the way in which they resemble Kant's schemata and at the same time the way in which they differ from Peirce's notion of icon as such:

---

26   On Peirce and diagrams see also Ketner (2023, 4–8).

A diagram, indeed, so far as it has a general signification, is not a pure icon; but in the middle part of our reasonings we forget that abstractness in great measure, and the diagram is for us the very thing. So in contemplating a painting, there is a moment when we lose the consciousness that it is not the thing, the distinction of the real and the copy disappears, and it is for the moment a pure dream – not any particular existence, and yet not general. At that moment we are contemplating an icon. (CP 3.362)

Differently from the diagram, the icon (similarly to the image in Kant) is not general, even though in the moment when we contemplate it as a thing in itself, neither is it particular. Therefore, it is a pure dream. In a similar manner, when contemplating visual diagrams, we have a tendency to cognitively associate the diagram with the thing it represents. A peptide chain represented as a necklace of concatenated amino acids is a diagram, in that it represents relations between every single amino acid in the chain, but there is the risk of exchanging the diagram for what it represents, with the result of contemplating an icon. In Peirce, diagrams are representations of Logic and Logic is part of Semeiotic, which penetrates the whole universe in its evolution. Thus, diagrams not only determine our reasoning and perceptions, being the mediators between categories and sensations; diagrams are a constitutive part of minds, and also quasi-minds. In the case of proteins as elementary life constituents, diagrams are schemata in the center of basic interactions guaranteeing the functioning of life. They relate things that have nothing in common. As categories and sensations are linked at the cognitive level, at the molecular level biomolecules of different nature are linked, such as proteins and their ligands: for a protein, a ligand is an external element having no direct relation to the protein. The only relation between a protein and its ligand exists thanks to the binding possibility: concave and convex shapes match each other. This kind of shape-matching, the basis of the binding of molecules, is a result of a diagrammatic relation between the two elements. It is general and not particular and it has an iconic nature because of a resemblance in structural relations but not in visual perception, therefore it is not a pure icon. The binding matching is what the protein shares with its ligand: it is a schema. By analogy, at the level of protein folding, the folding of a peptide chain into a functional protein also might be explained via schema/diagram. The complicated relation between linear peptide chain on the one hand and dimensional protein structure on the other hand is mediated through a schema, a diagram shared by the two elements in relation. The two are of very different order, one being linear and without biological function, the other being dimensional and functional. The two elements, so different yet so inseparable, share the substance of amino acids, which represents a diagram shared by both.

I already mentioned in what ways diagram differs from icon, mostly in its general characteristic and in that it is not necessarily related to a visual image. Additionally, we can specify that diagrams at the same time differ from pure icons. They are closer to symbols:

> It is true that ordinary Icons, – the only class of Signs that remains for necessary infer-
> ence, – merely suggest the possibility of that which they represent, being percepts minus
> the insistency and percussivity of percepts. In themselves, they are mere Semes, pred-
> icating of nothing, not even so much as interrogatively. It is, therefore, a very extraor-
> dinary feature of Diagrams that they show, – as literally show as a Percept shows the
> Perceptual Judgment to be true, – that a consequence does follow, and more marvel-
> ous yet, that it would follow under all varieties of circumstances accompanying the
> premises. It is not, however, the statical Diagram-icon that directly shows this; but the
> Diagram-icon having been constructed with an Intention, involving a Symbol of which
> it is the Interpretant […] It is the normal Logical effect; that is to say, it not only hap-
> pens in the cortex of the human brain, but must plainly happen in every Quasi-Mind in
> which Signs of all kinds have a vitality of their own […] The Schema sees, as we may say,
> that the transformate Diagram is substantially contained in the transformand Diagram,
> and in the significant features to it, regardless of the accidents […] The transformate
> Diagram is the Eventual, or Rational, Interpretant of the transformand Diagram, at the
> same time being a new Diagram of which the Initial Interpretant, or signification, is the
> Symbolic statement, or statement in general terms, of the Conclusion. By this labyrin-
> thine path, and no other, is it possible to attain to Evidence; and Evidence belongs to
> every Necessary Conclusion. (NEM IV: 316–19)

The aforementioned quotation contains a lot of information deserving a short commentary. Firstly, the nature of diagrams to represent a pure possibility or potentiality is crucial for our further argumentation. The possibility of what diagrams represent lies in the very nature of life: it is a might-be, they "merely suggest the possibility of that which they represent". Likewise, the information encoded diagrammatically in the genetic sequence is a mere possibility, genes might or might not be expressed under specific circumstances.

Secondly, the symbolic nature of diagrams is something that requires atten-
tion in order to distinguish them from other iconic semeioses as such. Since the diagram Peirce is referring to is a diagram constructed with a certain intention, it must necessarily be oriented towards a future, towards a possible future goal to be attained. This characteristic to be oriented towards a future can only be attained by a symbol, because of the fact that no icon nor index has the capacity to guarantee future representations; in other words, only the symbol is arbitrary and conventional enough to have the capacity to represent in future. Thus, also the genetic sequence has an inevitably symbolic character because otherwise it would be impossible to convey the unchanged information to future generations.

The arbitrary nature of diagrams with regard to their interpretation is another argument for diagrammatic symbolicity. In the words of F. Stjernfelt:

> A line may be interpreted in one diagram as a borderline, in another as a line of connection between two points, in yet another as a transport of some object between two locations. This may be banal, but nevertheless it is an important feature in the diagram's iconicity: the type only becomes apparent in light of the use of certain rules – long before the virtual application of the diagram on more specific meanings, not to talk about empirical reference. This implies that already the pure diagram is an icon governed by a rule, that is, by a symbol. (Stjernfelt 2007: 96)

Diagrams are not pure icons because pure icons are neither tokens nor types, limited to the category of firstness (Scott 2099: 121–123), but they have some indexical and also symbolic features. Scott focuses on diagrams in reasoning, talking about indexical or symbolic types of diagrammatic reasoning, but diagrams in molecular biology also indexical or symbolic; symbolic due to the future-oriented nature of the genetic code, and indexical because of the causal relations of interpreting the tracks of the environment or of previous generations in epigenetic terms.

Lastly, the quasi-mind is an important notion here, because it stresses the fact the diagrams exist not only in brains, but are somehow the constitutive principle of nature and the cosmos in the way they represent logic at the heart of semeiotic. The quasi-mind was explained in several places by Peirce, but let me quote this passage from Prolegomena to an Apology for Pragmaticism:

> Thought is not necessarily connected with a brain. It appears in the work of bees, of crystals, and throughout the purely physical world; and one can no more deny that it is really there, than that the colors, the shapes, etc., of objects are really there [...] Not only is thought in the organic world, but it develops there. But as there cannot be a General without Instances embodying it, so there cannot be thought without Signs. We must here give "Sign" a very wide sense, no doubt, but not too wide a sense to come within our definition. Admitting that connected Signs must have a Quasi-mind, it may further be declared that there can be no isolated sign. Moreover, signs require at least two Quasi-minds; a *Quasi-utterer* and a *Quasi-interpreter*; and although these two are at one (i.e., *are* one mind) in the sign itself, they must nevertheless be distinct. In the Sign they are, so to say, *welded*. Accordingly, it is not merely a fact of human Psychology, but a necessity of Logic, that every logical evolution of thought should be dialogic [...] what I have been saying is only to be applied to a slight determination of our system of diagrammatization, which it will only slightly affect; so that, should it be incorrect, the utmost *certain* effect will be a danger that our system *may* not represent every variety of non-human thought. (Peirce, Prolegomena to an Apology for Pragmaticism, in Bisanz 2009, 326)

To sum up, Peirce's version of Kantian schemata in the form of diagrams brings more light to the understanding of the semeiotic nature of elementary life structures, proteins. We might comprehend proteins as diagrams, thus specific forms of icons, icons which are general and potential, oriented towards the future and thus somehow teleological signs being produced and interpreted in the quasi-minds of nature. Proteins are iconic diagrams; this is given by their faculty to relate with totally different kinds of elements by resembling them thanks to their specific shapes (see more about the iconic nature of proteins in Faltýnek and Lacková 2021).

Of course, all life with proteins as one of its smallest elements is related to evolution, and thus to ever-changing forms of organic substance over time. For Kant (1781: 77), time is the basic schema of a pure concept, because time is but our product of understanding and interpreting the perceived world. For organisms, time is the elementary schema in the same measure as for human perception. Evolution is their way of interpreting the world where the interpretation of the world is at the same time living within it and dialoguing about it with past and future generations of species.

Diagrammatic logic

Peirce started his scientific career by studying chemistry in terms of the topology of relations. Especially diagrams, or schemata of molecular relations, were revolutionary at that time and became crucial elements in his later studies in logic (Ambrosio and Campbell 2017). The originality of Peirce's approach resides in his focus on observing molecular relations, rather than concrete chemical qualities; this approach was based on his interest in molecular topology and its spatial representations. Later applied to logic, Peirce's topological approach led to the creation of his well-known graphs. Peirce distinguishes two types of logical graphs: Existential and Entitative (MS 482, 59). I will focus exclusively on the Existential Graphs (EGs) because, besides their central importance for Peirce, the EGs have been widely extended, re-elaborated and reinterpreted in current philosophical research (Ketner et al. 2011, Bisanz et al. 2019, Pietarinen 2006, 2008, Pietarinen & Stjernfelt 2015, Shin 2002, Stjernfelt 2007, Zalamea 2017, Zeman 1964).

But apart from logic, the existential graphs are potent tools for the schematization of other sciences, thus having a real applicability in research: Beil and Ketner (2006) applies the diagrammatic notation to quantum physics, and Bisanz et al. apply them to neuronal activity (ISP 2019 in PS 10). In the following chapter, I design several examples of how a possible Beta version of the Existential

Graphs can be created for biosynthesis and related processes, translation, protein folding, enzymes, and gene expression in particular. In the words of Ketner:

> As such, it (semeiotic) can provide an insight into the methodological commonalities found in all science: physical, cultural, biological. Existential Graphs (EG) constitute the lingua franca for mature Semeiotic/Logic. The ontology required for EG is the same required for his Theory of Signs (or Semeiotic), namely: relations understood as the building blocks of reality. (Ketner 2011a: 376)

Peirce's work in logic consists mainly of an original elaboration of two kinds of notation of logical relations: the symbolic and the graphical. I will focus on the latter, since the graphical representations yield particularly constructive outcomes in biology, and also because of the current trend in topological explanations (Lacková and Zámečník 2020).

The EGs operate with basic notions such as relative terms, or *relatives*. Thus, the relational aspect of a graph is encompassed already in the terminology. The EGs are graphical diagrammatical systems depicting basic logical relations, and are threefold: Alpha, Beta and Gamma. While the Alpha and Beta graphs are perfectly elaborated systems, the Gamma graphs remain unfinished as graphical representations. In general, the Alpha graphs can be compared with propositional logic, the Beta graphs with predicate logic, and the Gamma graphs with several elements of modal logic. For the Alpha graphs, the primitive elements of the graphical logical system are symbols for sentences and cuts, which represent negation. The juxtaposition of graphs is interpreted as conjunction. For the Beta graphs, the primitive elements are symbols for predicates, cuts and lines of identity. According to Shin:

> The Beta system is analogous to a pure symbolic first-order language with an equality symbol, but without a constant symbol. (Shin 2002: 39)

In the Beta graphs, similarly to the Alpha graphs, the cut represents negation. The major difference between the Alpha and Beta graphs lies in the lines of identity (see Fig. 7). As in the case of the Alpha graphs, the cut represents negation, and therefore the main difference between the Alpha and Beta graphs is connected to the lines of identity expressing a relation of a shared variable for two or more predicates.

> A heavy line, called a line of identity, shall be a graph asserting the numerical identity of the individuals denoted by its two extremities (Peirce CP 4.444)

When Beta Graphs represent a connection of more than two predicates, we are dealing with multiple branching. The branching permits for a graphical representation much better than any symbolic notation; this is also why the Beta

```
  ┌─ is a bird
 ┌─── is black
  └─ is thievish
```

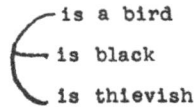

Figure 7: The identity line (CP 4.445).

Graphs are a suitable system of notation for the NonReduction Theorem, a theory based on Peirce's teridentity relation which I comment upon later in this chapter. As Shin has pointed out,

> this device, representing sameness iconically, makes a clear distinction between Beta graphs as a graphical system and other predicate languages as symbolic systems. While in a symbolic system tokens of the same type of letter represent the same individual that each token denotes, in Peirce's Beta system a network of lines represents the same individual denoted by each branch of the network. That is, Peirce's system graphically represents numerical identity with one connected network. (Shin 2002: 54)

Nonetheless, some researchers have tried to translate the Beta Graphs into symbolic logic, in an effort to demonstrate the alleged senselessness or vanity of graphical logic as such (Zeman 1964; Zeman 1986).

Logic of Relatives

Peirce's diagrammatic logic was not his first important achievement in logic: in the period when diagrammatic logic was first introduced to the scientific community, he was already known for his work in symbolic logic with innovations on notation, quantifiers, etc. His first attempt at a system of notation for diagrammatic logic was published in the 1897 in *The Monist* as a paper entitled The Logic of Relatives. The logic of relatives was already presented by Peirce already in 1870 in the essay Description of a notation for the logic of relatives, resulting from an amplification of the conceptions of Boole's calculus of logic, but still in the form of linear algebra. Peirce wrote several other papers during this period on the topic of the logic of relations, partially published in Volume III of the *Collected Papers* (for a minute introduction to the topic see Beil and Ketner 2012: 1957–1958). The diagrammatic representation of logic of relatives (1897) was probably the first case of putting into practice Peirce's diagrammatic logic. Even if the graphical representation from 1897 is not exactly the same system as Peirce's system of diagrammatic logic known as the EGs, it is in any case a sort of diagrammatic representation of logical relations even though not officially recognized as such. As Beil and Ketner point out, we have been unable to discover in his writings any overt instance in which he identified the types of diagrams which might apply to the logic of relatives and its associative algebra. This omission is surprising considering that both diagrams and algebras occupied a large part of his time. The omission is also surprising since there is, indeed, a set of diagrams which does correspond to the algebra of the logic of relations. These diagrams are the beta existential graphs […], in particular, the triadic subset of these graphs. (Beil and Ketner 2012: 1968)

Logic of relatives uses the concept of relative terms, or relatives:

> A *relative* is a term whose definition describes what sort of a system of objects that is whose first member (which is termed the *relate*) is denoted by the term; and names for the other members of the system (which are termed the *correlates*) are usually appended to limit the denotation still further. In these systems the order of the members is essential; so that (A, B, C) and (A, C, B) are different systems. As an example of a relative, take "buyer of – for – from"; we may append to this three correlates, thus, "buyer of every horse of a certain description in the market for a good price from its owner."
> A relative of only one correlate, so that the system it supposes is a pair, may be called a *dual* relative; a relative of more than one correlate may be called *plural*. A nonrelative term may be called a term of *singular reference*. (CP 3.218. – 3.219)

To illustrate the concept of the logic of relatives let's start with a short passage from the 1897 essay:

> An essential part of speech, the Preposition, exists for the purpose of expressing relations. Essential it is, in that no language can exist without prepositions, either as separate words placed before or after their objects, as case-declensions, as syntactical arrangements of words, or some equivalent forms. Such words as "brother," "slayer," "at the time," "alongside," "not," "characteristic property" are relational words, or *relatives*, in this sense, that each of them becomes *a general name when another general name is affixed to it as object*. In the Indo-European languages, in Greek, for example, the so-called genitive case (an inapt phrase like most of the terminology of grammar) is, very roughly speaking, the form most proper to the attached name. By such attachments, we get such names as "brother of Napoleon," "slayer of giants," "{epi 'Ellissaiou}, at the time of Elias," "{para allélön}, alongside of each other," "not guilty," "a characteristic property of gallium." *Not* is a relative because it means "other than"; scarcely, though a relational word of highly complex meaning, is not a relative. It has, however, to be treated in the logic of relatives. (Peirce, The Logic of Relatives, in Bisanz 2009, 187)

As is evident from the quote, the logic of relatives is to be initially a propositional logic, since all propositions in language are relations. More concretely, the best illustration of a relative in language is a preposition. Yet relational logic for Peirce becomes necessarily triadic, and at that point the preposition is no longer a suitable example of a triadic relation. The best example of a triadic relation is a verb:

> A verb by itself signifies a mere dream, an imagination unattached to any particular occasion. It calls up in the mind an *icon*. A *relative* is just that, an icon, or image, without attachments to experience, without "a local habitation and a name," but with indications of the need of such attachments. (Peirce, The Logic of Relatives, in Bisanz 2009, 187)

A relative is in Peirce's words "a mere dream" to which correlates must attach to complete "its mode of existence of a haecceity" (CP 3.460). Peirce's idea of relatives was inspired by verbal valency: verbal arguments are correlates which complete the pragmatic meaning of the verb. Because of this he is also in some sense a precursor of Tesnière's valency theory in linguistics (Paolucci 2006).

The valental aspect of graphs is described in detail in MS 482, titled On Logical Graphs:

> A valental graph is a graph of which every spot, by virtue of its color, or inherent quality, has a determinate number of bonds attached to it on determinate sides of it. An attachment to one side of a spot might be different in Kind from an attachment to another side. (MS 482: 2)

Valency analysis is what lead to Peirce's diagrams in the most general way, and to his Existential Graphs:

> Indeed, application of the technique of classifying and analyzing in terms of Valency Analysis, named the doctrine of Cenopythagoreanism, created a number of other results within Peirce's whole system, for example: the Existential Graph method of logical diagrammatization [...] the doctrine of the categories which is a central part of Phanerocsopy, and the classification of signs. (Ketner 1986: 12–13, cited from Scott 2009: 86).

Valency, diagrams and relational thinking are all important notions for the description of the phaneron. According to Peirce, the Existential Graph "represents the structure of the phaneron to be quite like that of a chemical compound" (NEM IV, 320). Phaneron is defined by Peirce as "collective total that is present to the mind":

> By the *phaneron* I mean the collective total of all that is in any way or in any sense present to the mind, quite regardless of whether it corresponds to any real thing or not. If you ask present *when*, and to *whose* mind, I reply that I leave these questions unanswered, never having entertained a doubt that those features of the phaneron that I have found in my mind are present at all times and to all minds. So far as I have developed this science of phaneroscopy, it is occupied with the formal elements of the phaneron. (MS 1334, 35–36)

Indecomposable elements of the phaneron are represented as ordinary spots on the diagram, but the chemical analogy does not stop here. Peirce extends it further to biology, concretely to the proteins:

Proteid Analogy: but whatever I may conjecture as to those vast super-molecules, some containing fifteen thousand molecules, whether it seems probable on chemical grounds, or not, that they contain groups of opposite polarity from the residues outside those groups, and whether or not similar polar submolecules appear within the complex inorganic acids, it is certainly too early to take those into account in helping the exposition of the constitution of the phaneron. (NEM IV, 321)

Of all of Peirce's reasoning in all the sciences, his diagrammatical reasoning is what best characterizes his scientific method: relational thinking is applicable in all sciences and valency analysis is a universal scientific tool. In Ketner's words, diagrammatical reasoning is not different from mathematical reasoning (Ketner 1986: 12–13, cited from Scott 2009: 86).

But what need was there actually to create a diagrammatic system for the logic of relatives if a symbolic notation already existed? We must start our argumentation with the presupposition that, for Peirce, every reasoning is only by semeioses and every semeioses is triadic. Therefore, also in logic there is no other way how to represent basic logical relations than the triadic one, so the primitive relative term is a triad, and an example of this would be the trivalent verb "to give". Now, as was stated above, a relative is only completed if all its blank spaces (blanks) for attachments are occupied, ergo we need three correlates, three arguments to complete the meaning of the verb "to give". Now, if we linked up these three arguments one after another with a kind of linear symbolic notation, it would be impossible to represent accurately the uniqueness of a triad. Peirce came to his discovery thanks to analogical diagrammatic notation in chemistry (see Fig. 8):

A chemical atom is quite like a relative in having a definite number of loose ends or 'unsaturated bonds,' corresponding to the blanks of the relative. (*The Logic of Relatives* in Bisanz [ed] *2009*, 191, originally published in *The Monist*, Volume 7: January, 1897: 161–217)[27]

A chemical molecule consists of chemical atoms, and the manner in which atoms are connected with one another is based on the number of loose ends of each

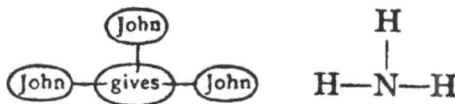

Figure 8: A chemical atom is quite like a relative.

---

27   On valental aspect of the verb "to give" see also MS 482, 7.

atom. We can observe the formula of the ammonia molecule in the figure: three atoms of hydrogen bind to potentially loose ends (blanks) of the atom of nitrogen. By analogy, the verb "to give" in the sentence "John gives John to John" has potentially three blanks which are occupied by co-relates "John".

In chemistry the source of the valency analysis is mathematical topology (Scott 2009: 36). Peirce explains this diagrammatic representation of the verb "to give" as follows:

> Is relation anything more than a connexion between two things? For example, can we not state that A gives B to C without using any other relational phrase than that one thing is connected with another? Let us try. We have the general idea of *giving*. Connected with it are the general ideas of *giver*, *gift*, and "donée." We have also a particular transaction connected with no general idea except through that of giving. We have a first party connected with this transaction and also with the general idea of giver. We have a second party connected with that transaction, and also with the general idea of "donée." We have a subject connected with that transaction and also with the general idea of gift. A is the only hecceity directly connected with the first party; C is the only hecceity directly connected with the second party, B is the only hecceity directly connected with the subject. Does not this long statement amount to this, that A gives B to C? (Peirce, The Logic of Relatives, in Bisanz 2009, 188)

The idea of giving cannot be reduced to dyadic relations between the three arguments – it is a unique act irreducible to any kind of pair relations concatenated one after another. This example illustrates a fundamental principle of the NonReduction Theorem.

NonReduction Theorem

Although Peirce never explicitly formulated the so-called NonReduction Theorem or Reduction Thesis, Burch (1992) formulated it and provided a mathematical demonstration of the theorem, now many Peircean scholars widely use this term (Burch 1997; Ketner 2011b). Peirce has nevertheless clearly defended the idea of (non) reduction in a number of places in his work. Consider the following:

> For were every element of the phaneron, a monad or a dyad, without the relative of teridentity (which is, of course, a triad), it is evident that no triad could ever be built up. Now the relation of every sign to its object and interpretant is plainly a triad. A triad might be built up of pentads or of any higher perissid elements in many ways. But it can be proved – and really with extreme simplicity, though the statement of the general proof is confusing – that no element can have a higher valency than three. (CP 1.292)[28]

---

28   Certain authors (Brunning 1997: 252) mention Peirce's correspondence with Lady Welby (SS:43) where the NonReduction Theorem is expressed very clearly: "I prove

The idea of the non-reduction of triadic relations is central to all of Peirce's thinking. This is probably why Peirce himself never formulated it as a theorem, since the nonreduction of any triadic relations of any form is implicit throughout his entire work. Nevertheless, non-reduction is also stated explicitly in some manuscripts (see the quotations above) and is exhaustively explicated in the logic of relatives in terms of the distinction between genuine and degenerate triads (Peirce in Bisanz 2009, 186–229).

The essence and uniqueness of Peirce's teridenty relation resides in its ontological trifold, status which goes beyond mere juxtaposition of the three elements. The mode of being of teridentity is not reducible to linear spatial arrangement. The limits of juxtaposition are described in Brunning (1997).

The NonReduction Theorem (NRT) is often associated with the Relational Completeness Theorem (RCT). Indeed, many authors conflate the two theorems, which is explained by the two theorems' interconnection, yet putative interdependence. Ketner formulates both theorems as follows:

> (I) Triadic relational types cannot be constructed using bonding and a resource base consisting only of monadic and/or dyadic relational types (NRT);

and

> (II) Relational types of any valency can be constructed using bonding and a resource base consisting of monadic, dyadic, and triadic relational types (RCT). (Ketner 2011b, 9)

Burch, in contrast, subsumes both theorems in one thesis, the so-called Reduction Thesis, in which NRT and RCT are present in the form of positive and negative components.

> Peirce's Reduction Thesis is a doctrine that has both a negative and a positive component. The negative component of the Thesis says, first, that relations of adicity 2 may not in general be constructed from relations exclusively of adicity 1; and, second, that relations of adicity 3 or greater may not in general be constructed from relations exclusively of adicities 1 and/or 2. The positive component of the Thesis says that all relations, regardless of the domain of arbitrary (non-negative integer) adicities may be constructed from relations exclusively of adicities I, 2, and 3. (Burch 1992: 670)

The RCT concept suffuses all of Peirce's thinking. Apart from its connection to the NonReduction Theorem, relationality in general constitutes the core of all of Peirce's philosophy. The theory of continuum (synechism) is a concept

---

absolutely that all systems of more than three elements are reducible to compounds of triads […] The point is that triads evidently cannot be so reduced." This is Peirce's letter to Welby from December 13th, 1904.

encompassing both Peircean semeiotic and mathematics and is closely related to the topological understanding of geometric relations.

Diagrammatic Formalization of Protein Biosynthesis and Protein Structures

Diagrammatic logic permits us to express and graphically convey the meaning of complex relations such as triads and polyads. Beil and Ketner have already shown the possibility of using this instrument for quantum mechanics: graphical logical representation has several advantages when compared to linear notation. I would like to demonstrate that diagrammatic logic is also potentially a useful tool for molecular biology. We mentioned genotype-phenotype maps as examples of topological thinking in biology. As will be shown, not only the relation between genotype and phenotype, but many other types of relations in biology can profit from a topological or diagrammatic approach. To start with, I propose to see applications of both the identity line from the Beta Graphs and the graphical notation from 1897 to elementary molecular processes at the level of protein synthesis, which I introduced earlier.

The translation from RNA transcript to the peptide chain requires the use of a code. The translation is not based on chemical affinity or physical forces between bases and amino acids, therefore the code mediates between the two elements and tRNA is the physical mediator. The relation between triplets of bases and amino acids is not dyadic because they never have direct interaction. More importantly, there must be a third element which is a completely abstract one, and it is the code. Without the code, the matching would be simply random, but in nature the matching between triplets and amino acids is not random, rather corresponding to the very precise rules of genetic code see attached genetic code table. This code can be understood as interpretant I within an O, R, I semeiosis, where the representamen is the RNA and object is the peptide chain. To illustrate this very complicated relation – which is unique since the process happens in a very small amount of time and is not sequential – we can use the notation from 1897 inspired by the chemical molecule, and which is also applied to verb valency (see Fig. 9, compare with Fig. 8):

As a second example of the application of diagrams in molecular biology I take the case of protein folding. Using the same sample from 1897, the relation between sequence and structure might be illustrated as follows in Fig. 10, alternatively with line of identity, as illustrated in Fig. 11.

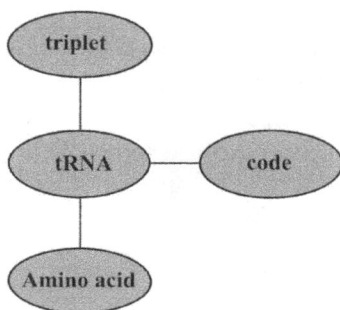

Figure 9: The valency of DNA translation.

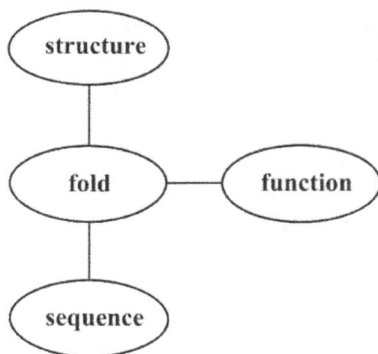

Figure 10: Sequence and protein structure relation.

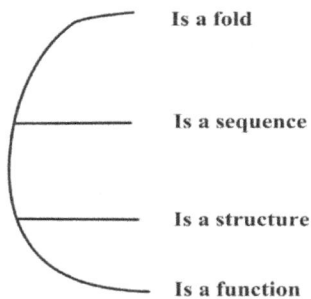

Figure 11: Identity line in protein folding.

The sequence gives structure to the protein by folding itself thus the fold belongs at the same time to both structure and sequence, but this is not the most interesting point about protein folding. The conditions and factors which enable a peptide chain to fold are multiple. The folding depends on the hydrogen bonds, hydrophobic interactions, van der Waals forces, thermodynamics, temperature and many other factors. The folding happens with help of ribosome molecules and also chaperone molecules. These are often called as "helpers" as if they were actual agents helping the protein fiber to fold. What makes protein folding fascinating is the fact that the way the peptide chain is folded gives the protein its function. In Figure 10, the identity line connects three predicates related to the protein fold (compare with Fig. 6). The particularity of biology is that it is impossible to avoid involving abstract concepts such as *function* or *code*, and this also explains the attraction of teleological explanations or vitalism and similar theories. I hope that diagrammatic logical notation might at least provide an effective instrument to represent the uniqueness and inseparability of real but non-material concepts connected to life and its routine relational processes. In the same way that Peirce positioned himself somewhere in between idealists and materialists – or better yet, he was a kind of materialistic idealist in the sense of the assumption that mind and matter evolve in parallel (Brier 2019) – molecular biology (and biology in general) is irreducible to physics or chemistry understood in the physicalist manner. It is possible to express such statements without being considered a neovitalist.

Chapter 5   Beyond Beta: Gamma Graphs and Synechism in Biology

Singularity and Potentiality

So far, various mathematical achievements by Peirce have been commented upon in connection with their possible application to biology, mostly as presented in the book by F. Scott (2006) and in the essay by Ketner and Putnam, *The Consequences of Mathematics* (1992). The notions of diagrams and topology have also been introduced. Potentiality was also already mentioned several times, yet dissociated from mathematics. Potentiality, however, is a central concept for Peirce's mathematics, mostly in the theory of infinitesimals. For Peirce potentiality was more than a mathematical construct, and has become a part of his metaphysics, especially within the theory of synechism. I will proceed further with argumentation about the potentiality of an infinite number of biological structures and relations compared with actually existing structures in nature, demonstrating that potentiality is something not only in connection with evolution. I will be chiefly interested in protein folding and the notion of potentiality in creating protein structures, but also in other domains of molecular biology related to protein biosynthesis.

My contribution in this chapter aligns with the project called Biology of Mind at the Institute for Studies in Pragmaticism in Lubbock (Bisanz et al. 2019), which consists in the application of Peirce's relational logic to biological and neural phenomena. The Existential Graphs, mostly the Beta Graphs, have been applied to recent studies in neuroscience by Bisanz and Ketner. The Biology of Mind project is only one part of the Interdisciplinary Seminar on Peirce, which is organized by the research group at Texas Tech University in Lubbock (ISP)[29]; its main focus is to demonstrate the relevance of Peirce's relational triadic logic to contemporary science. The consequence of this endeavor is more far reaching than Peirce's ideas themselves. The ISP project can also be understood as a position statement regarding questions like *What is the correct reading of Peirce,* and debates surrounding the idealist versus the realist reading of Peirce. It has been suggested that neither of the labels of realism or idealism adequately describe the scope of Peirce's work (see e.g. Hausman 1993: 140–190; Nöth 2014).

---

29   Elize Bisanz, Kenneth Ketner, Clyde Hendrick, Levi Johnson, Michael O'Boyle and others.

The ISP project resolves the idealistic/realistic dilemma with the Objective method, consisting in hypotheses, laboratory experiments and their formalization in diagrams, the Existential Graphs. The Existential Graphs are existential because they refer to "mathematical and logical analyses of relations as they exist [as they are discoverable realities within] scientifically observable phenomena" (Bisanz et al. 2019: 20). Thus, the example of a diamond and its quality of hardness from *How to make our ideas clear* from 1878 acquires a new meaning when we consider potentialities as realities in nature in terms of the manuscript *The Law of Nature*. Peirce's triadic relational logic transcends the idealistic/realistic dilemma in this way.

Paolucci has also questioned Peirce's presumed "return to realism" after 1885, as it is described by Murphey (1961) and developed by Fumagalli (1995) and Short (1996). Paolucci reminds us that it was precisely around the year 1885 when Peirce was working on the finalization of his logic of relatives, which is by definition incompatible with naïve realism (Paolucci 2015). Naïve realism is dyadic by nature, while Peirce's logic and semeiotic is defined by triads as the primary relation. Peirce's notion of growth, from the manuscript *Law of Nature* analyzed in the first chapter of this book, asserts the future realities of germ cells, eggs and seeds. Empirical evidence provided by genetics and epigenetics also supports an argument against naïve realism in terms of real possibilities discoverable in nature. They exist in the tension between the digital script and its future embodiment in matter. This is how Peirce's position should be understood in light of the current scientific paradigm. The objective method is a result of understanding Peirce's metaphysical position as a *relational realist* (Ketner 1999). This simply means that relations are real, thus semeioses are real even if non-material. Thus, for example, (non-material) communication/dialogue between scientists is part of laboratory science (Ketner 1999: 25). The Objective method is inclusive and non-reductionist, one consequence of which is that humanities and the arts are equally classified as sciences according to the objective method (Ketner in Bisanz 2009: 56, see also Ketner 2023: 3).

> If we destroy the purported equivalency of science with scientism, we find that science is broader and more inclusive [...] Science understood as objective method can inquire into any area of human interest in which realities might be present whether as existent or reals waiting for an open mind to discover them. (Ketner 1999: 27)

When I apply the objective method to molecular biology and protein folding, I am dealing with both existent and non-existent realities. As will be explained, the conception of potentiality in molecular biology is connected to strings, whether they be RNA strings, DNA strings or peptide chains. From a reductionist perspective it can be said that all basic life structures are strings: all life is hidden in the strings of nucleic bases or in the strings of amino acids. This means that life is virtual if we consider that it is encoded by genetic information in the form of linear virtual genetic information. This way of understanding life is partially derived from the language metaphor of the genetic text (Jakobson 1971), which will be discussed later in greater detail, but let's consider to what extent this metaphor holds in terms of virtuality.

First of all, the difference should be distinguished between what is potential and what is virtual. As a matter of fact, we can't really state that the information encoded in DNA is virtual in the same sense that information encoded in human language is virtual: the difference between language/writing systems and DNA strings lies in the fact that DNA macromolecules are materially existing entities, therefore even if the information they encode might be hypothetically virtual, the very medium or "storage" is a material organic molecule. DNA is often referred to as a "virtual script" in opposition to genes expressed in living bodies in the form of specific features, and to illustrate the genotype-phenotype distinction. But still, the genome itself, even for unexpressed genes, is materially present in the nuclei of our cells. Therefore, it is somehow there at the molecular level even if unexpressed at the level of protein biosynthesis. To explain this complex relation between genes and their expressions in the form of proteins or higher bodily structures, it is opportune to use the notion of potentiality. Potentiality will serve in my further argumentation about the process of protein folding, where all peptide strings are potentially folded.

So far in this book, the way of understanding proteins as products of biosynthesis has always been treated as unidirectional, which means from sequence to structure (the sequence of amino acids leads to the structure of the protein). In other words, the linear peptide chain has been comprehended as an entity formally and substantially determining the final protein structure. "Formally" because the sequence order of singular amino acids is supposed to encode the final folds, to determine where folding is to occur within the positions of the chain; "substantially" because it is the peptide chain itself from which the protein is made. As a consequence, the peptide chain has been described as an entity already articulated when entering into relation with the protein's shape and function. It is articulated linearly as a sequence of amino acids concatenated one after another and coded in the RNA transcript. To be precise, when I say that organic matter such as peptide chain is linear, what I am saying is that it is composed of individual units lined up one after another. To put it a different way, to be linear presupposes the existence of individual entities, distinct and separate individual features composing the entity that is to be linear. What is surprising is that not only biologists (Monod 1972; Jacob 1970) but also philosophers (Deleuze and Guattari 1987) and linguists (Jakobson 1971) treat the peptide chain as an entity pre-structured and articulated *per se*. I hold the opinion that there is another option for describing protein fiber. I propose a small thought experiment. Let's suppose the peptide chain to be an unarticulated mass, "a body without organs" (Deleuze and Guattari 1987), an amorphous continuum, emancipated from the familiar image of a protein fiber as a chain of beads on a necklace. After all, a necklace is nothing but the habit of scientific schematization of biological entities which could be schematized in a different way, for instance as a simple continuous line.

Recall that there is divergence between scientific models and what they actually represent (similarly to the problem of contemplating an icon). Sometimes it is hard to differentiate between scientific models and reality (Markoš and Švorcová 2009), yet one should keep in mind that scientific models are a matter of vogue in the contemporary scientific paradigm. I did not use the term "vogue" accidentally; sometimes scientific models resemble real artistic masterpieces and correspond with the scientist's aesthetic taste (e.g. Haeckel's famous biological drawings). But, let's go back to our argumentation. To imagine a peptide chain in a non-linear way means, without particular individual amino acids composing it, one simply has to comprehend the chain in a molar and not a molecular way. The molecular represents the distinctions between singular points, which are irrelevant for the interpretive process. Molar on the other hand represent only the functional differences responsible for functional semeiosis. The terms molar and molecular originate in Deleuze and Guattari (1987) who conceived of the molar as that which is conceivable and relevant, while the molecular is that which exists but is not relevant (for the difference between the concepts of molar and molecular see also Eco 2007). This model, the "molar peptide model", liberates us from the pre-established sequentiality of the peptide chain. The sequentiality of the amino acid chain is its characteristic feature, and there is a correspondence between amino acid sequences and particular protein folds, as demonstrated by Chothia and Lesk (1986).

Much like language, where no content is ever perfectly expressed in linear strings of words, neither the protein structures nor their functions can be exhaustively transcribed into linear strings. One dimension is always missing. Linear strings are the perfect means of economization for the sake of conserving information, but economization inevitably entails the loss of some important content.

Before I proceed with the thought experiment, let me comment on the notion of arbitrariness in protein folding. It can be said that the relationship between the protein fiber and the function of a protein is also a matter of arbitrariness: in other words, it is a result of evolutionary convention and not a result of physical necessity. Of course, arbitrariness by itself is not a constitutive principle of a protein folding "code". Quite the contrary, physical laws of thermal stability determine the structure of a folded peptide chain, but there remain important features of protein folding that are definitely arbitrary, starting with the choice of the twenty amino acids and ending with Levinthal's paradox. In order to better anchor the understanding of the peptide chain in terms of arbitrary syntactic rules, I turn to J. Monod:

> With the globular protein we already have, at the molecular level, a veritable machine – a machine in its functional properties, but not, we now see in its fundamental structure, where nothing but the play of blind combinations can be discerned. Randomness caught on the wing, preserved, reproduced by the machinery of invariance and thus converted into order, rule, and necessity. A totally blind process can by definition lead to anything: it can even lead to vision itself. In the ontogenesis of a functional protein are reflected the origin and descent of the whole biosphere. And the ultimate source of the project that living beings represent, pursue and accomplish is revealed in this message – in this neat, exact but essentially indecipherable text that primary structures constitute. Indecipherable, since before expressing the physiologically necessary function, which it performs spontaneously, in its basic make-up it discloses nothing other than the pure randomness of its origin. (Monod 1972: 98)

What Monod describes in this part of his famous book is in fact nothing more than the difference between "s-codes" and "codes" (Eco 1984), and the need to comprehend expression as inherent in content (meaning). In other words, for Monod the "indecipherable text" of protein fiber represents expression, the "necessary function" represents content, and "randomness" is nothing but another term for arbitrariness. I consider the aforementioned quotation by Monod to be essential, since it is one of the rare examples of semeiotic thinking in biology: the inseparability of expression from content (and indecipherability of expression by itself) as one of the basic characteristics of protein code. Before expressing the protein function, the peptide chain is superfluous, redundant – it is as if there was no sense in its existence without its relation to protein structure. In other words, there would be conceivably no peptide chains in nature if there were no proteins. Thus, it is absurd to comprehend peptide chains as something that predetermines protein functions. Language metaphors in this place may serve to better understand the line of reasoning. In natural language, singular phonemes do not preexist language itself: they exist potentially in the physical world of sounds but they do not possess their particular identity as phonemes, the smallest linguistic units. The unique difference between language units and biological units is in materiality, the genetic material (even if potential and unexpressed) is inherent in every cell of an organism, therefore it is not virtual in the same way as language is.

Now let's proceed with our thought experiment. Suppose that the sequentiality of amino acids within a peptide chain does not generate the protein structure, but, conversely, it exists only thanks to and because of the protein structure: "indecipherable before expressing the function" means that before the birth of a functional protein the expression itself has nothing to encode and thus, we cannot even consider it as having semeiotic reality. That means that the individual existence of single amino acids emerges only thanks to protein structure: thanks to the folds that give birth to the structure. Until the moment of folding, the peptide chain is, let's say, an amorphous continuum.

A single peptide chain represents an inconceivably enormous amount of possibilities of potential structures (remember Levinthal's paradox: at least 5×10^{47} possibilities and probably many more). When one sees a peptide chain, they see nothing other than an "infinity" of possible structures. Thus a chain is nothing more than mere possibility (firstness) without concrete identity (structure) and is potentially constitutive of an almost infinite amount of possible structures (Levinthal's paradox). For this reason, it cannot be said that sequence determines structure, since sequence determines a potentially infinite number of structures. We should say, on the contrary, that structure determines sequence, as far as only when the structure is given can the sequence adopt its identity. The amino acid residues of the final fold are differentiated as binding places (relates, if we use the terminology from the logic of relatives), thus, we can say that amino acids adopt their identity. Compare this with the following passage by Peirce:

> When we say that of all possible throws of a pair of dice one thirty-sixth part will show sixes, the collection of possible throws which have not been made is a collection of which the individual units have no distinct identity. It is impossible so to designate a single one of those possible throws that have not been thrown that the designation shall be applicable to only one definite possible throw; and this impossibility does not spring from any incapacity of ours, but from the fact that in their own nature those throws are not individually distinct. The possible is necessarily general; and no amount of general specification can reduce a general class of possibilities to an individual case. It is only actuality, the force of existence, which bursts the fluidity of the general and produces a discrete unit. (CP 4.172)

Only actuality produces discrete units, and only protein structure produces amino acids as discrete units of a particular fold. Only when "the die is cast" can the throw acquire its particular reality, its identity, and until that moment it is only a possibility. So to speak, a peptide chain is real only as a potentiality, and its particular amino acids are also real only as potentiality. Unless they are folded, they are not real singularities, which is why proceeding from sequence to structure does not give desirable results in predicting protein structures. We can apply this optic also to evolution – many potential organismic forms have existed in the form of potentialities, but only some of them happened to exist as real singularities.

In Chapter 3 the notion of real possibilities was introduced as it relates to epigenetics. Protein folding corresponds even more intrinsically to real possibilities. Potentiality is the clue for understanding the relation between the linearity and dimensionality of the peptide chain. It is the idea of the potential as firstness, a mere would-be as an Icon, and not the idea of the rule as Thirdness or symbol, that expresses a more complex way of describing the basic building blocks of life, which are proteins or proteic structures. The peptide chain is only potentially folded, but this potentiality is real in the sense that it is experimentally proven that most parts of peptide chains will fold in natural conditions in order to form functional proteins.

An important note to make at this point is that the folding of the peptide chain is potential and at the same time is not a rule; experimental evidence can be found for this statement. I can refer to the class of proteic structures, so-called intrinsically disordered proteins (IDPs), also known as intrinsically unstructured proteins (Wright and Dyson 1999; Dyson and Wright 2005). This class of proteic structures are peptide sequences with unstructured regions at least 50 amino acids long. The word "unstructured" means that they are simply unfolded chains or are not associated with strictly stable folding. The particularity of this group of proteins resides in that, even if they lack an inherent structure, they are functional and they behave as naturally structured proteins, usually performing the following functions: DNA recognition, transcriptional activation, translation, membrane transport, cellular signal transduction and many others. Some IDPs are induced to the fold by interactions with other molecules, and these proteins have several important advantages in processes involved in cellular signaling and regulation. Unstructured proteins are inherently flexible and their structure can be easily shaped by their proximate environment. Many IDPs are able to acquire a temporarily stable structure when in contact with their target molecule (Wright and Dyson 1999: 321). A target molecule not only shows the capacity of categorization of IDPs (the choice of one particular target molecule from a plurality of options), but additionally supports the stability of the protein structure. IDPs are thus an even better example of existence on the edge between potentiality and actuality.

The existence of a group of proteins which do not have an intrinsic structure but do have a function contradicts the central dogma of molecular biology,
which holds that there is only one direction going from sequence to structure
(and consequently to function). IDPs disprove the one-to-one correspondence
between structure and function in proteins, even though the research trying to
find regularities in the relation between the sequence of IDPs, their conforma-
tional properties, and protein function, is growing: sequential properties of IDPs
are currently under scientific investigation (Das et al. 2015).

For the purpose of my argumentation, IDPs are an example of real possibilities in nature and proof of the fact that protein folding is potential and at the
same time is not a rule. In the case of IDPs, the prediction of protein structures
is indeed more ambitious than in the case of ordered proteins. Even if disordered
proteins do not have an intrinsically stable structure, they might acquire temporarily stable structure in order to perform a concrete biological function, and
this is fairly common. Thus, IDPs are a perfect example of the combination of
chance, law and habit, a natural example of evolutionary Love exercising its force
of attraction to fold the unordered chain for the sake of biological function[30].

Potentiality combined with Law, in other words Logic, is at the heart of nature,
and biological research has borne out this fact since the first works on evolution,
but surely even long before that.

> Real possibilities thus connect epistemology, expressed in the pragmatic maxim,
> to ontology: real possibilities are what science may grasp in conditional hypotheses.
> (Stjernfelt 2007: 40)

To comprehend this kind of Logic, a logic which governs biological structures
and allows for real possibilities, in a scientifically formalized manner, a novel
kind of logical notation is needed. Diagrams prove to be a plausible proposal for
this enterprise.

This also explains why Peirce preferred diagrams as the notational tool for his
logic. Logic is for Peirce in continuity with Semeiotic and Evolution, thus it must
also reflect the potentiality of biological phenomena. Space and time cannot be
dissociated from the notion of potentiality. Diagrams and icons are also potential
insofar as they represent their objects through similarity. Similitude is potential
because it requires the interpreter/observer. There must be someone to decide
upon the similarity (Eco 1984). This is not the case with symbols, since they

---

30   Theoretical biologist Švorcová even speculates that IDPs are primarily compared to
     minor order proteins in a cell, and the impression that ordered proteins are in the
     majority is only a reflection of this laboratory method (personal communication).

are based on general rules independent from singular criteria of similarity. But diagrams are a special kind of icon, as was discussed earlier, having a symbolic nature in that they refer to the future and do not represent visual but structural iconicity. Even more radically, the argumentation can go in the direction of proclaiming that it is the potentiality in the form of diagrams which is intrinsic in nature. The statement that diagrams are somewhat inherent in nature is verbalized in the following quote by Peirce:

> But we do not make a diagram simply to represent the relation of killer to killed, though it would not be impossible to represent this relation in a Graph-Instance; and the reason why we do not is that there is little or nothing in that relation that is rationally comprehensible. It is known as a fact, and that is all. I believe I may venture to affirm that an intelligible relation, that is, a relation of thought, is created only by the act of representing it. I do not mean to say that if we should some day find out the metaphysical nature of the relation of killing, that intelligible relation would thereby be created. No, for the intelligible relation has been signified, though not read by man, since the first killing was done, if not long before. (NEM IV: 316)

This is a very significant passage from NEM where Peirce questions the metaphysical nature of diagrams in regard to the relation between a diagram as a representamen and the object it represents. A diagram as a representamen is necessarily visual or of another sensory nature and it represents some object to which it is related by the relation of mediation of a schema in Kant's sense of the word. The intelligible relation signified, though not read by man, would be the relation of the schema; we can imagine it as Kant's transcendental schema, which exists independently from any human cognition about the act of killing, but exists in the very nature of the act of killing.

Since the birth of modern science, iconicity became unscientific due to the growing importance of symbolic formalism not only in science but also in humanities, predominantly in linguistics and semeiotic in the form of structuralism or generativism. But Peirce demonstrates how symbolic formalism is not the only possible way of doing science. Since the very objects of study of science prove to be potential, as in modern physics and biology, but also in other branches of science, also the notation system for science might be based on diagrammatic notation that includes potentiality. This does not mean, however, that diagrams and their potentiality as Peirce elaborates them, are acceptable without problems by the current scientific community. Surely Peirce's doctrine on diagrams, if we embrace it in its complexity, can present several problems, not all of them admissible by the prevailing scientific paradigm. As Stjernfelt points out,

Peirce's doctrine of would-bes nevertheless contains numerous problems. Being prime examples of Thirdness, they embody real existent continuity – but by the same token there is a tendency in Peirce to let real possibilities incarnate all the very different metaphysical issues which the category of Thirdness is expected to solve. This includes no less than habit, symbols, teleology, mind, purpose, evolution, life; sometimes even personality, love, God, etc. (Stjernfelt 2007: 42)

The notion of real possibilities will be once again developed, this time diagrammatically and in the context of protein folding: genetic information exists in the nucleus in a digital form unless the folded protein is materialized into physical existence. Until that point it only exists as a potentiality, as a non-existent real (Ketner 2011a: 383–384). This is to demonstrate that non-existing realities are part of nature. I will treat all the related metaphysical implications not as "problems", but as parts of nature and inherent characteristics of biological organisms.

Formalisation of Potentiality in Biology: Selected Examples

Now I will discuss the formalization of such relations of potentiality in terms of Peirce's Beta Existential Graphs, using the concept of stems, or "potential hooks", as proposed by Bisanz et al. (2019: 35–47). The potential hook is a concept added to the original Beta version of Peirce's EGs) in order to enable a more formal and diagrammatic description of potential relations. To begin, I summarize the system of Betagraphic as proposed by Bisanz et al. (2019), but elaborated already in ISP 2011 and ISP 2015, and additionally I apply this graphical system to molecular biology.

The Betagraphic by Bisanz et al. operate with relations between variables and constants, which can be of three basic atomic kinds: unary, binary and ternary. These three types of Beta relations eventually correspond to Peirce's monads, dyads and triads. Relations are created by the operation of *bonding*. For the diagrammatic representation of relations, the authors use spots and horizontal lines, also called *hooks* (Bisanz et al. 2019: 22). Spots represent relations themselves while hooks are labeled by letters denoting variables. The three atomic relations might be schematized as in the figure below. The so-called "molecule" (ISP 2015) in the Beta Graphs is a finite sequence made of bonds between various separate graphs or "atoms".

In Fig. 12 we can see in the upper line the three atomic relations (monadic, dyadic and triadic) and in the lower line the so-called molecule (ISP 2015). Bisanz et al. (ISP 2019 in (PS 10)) propose a number of useful logical operations to apply to this modified version of the original Beta Graphs that permits for interesting interdisciplinary purposes, such as: blocking, potential hooks/potentials stems, and branch generation.

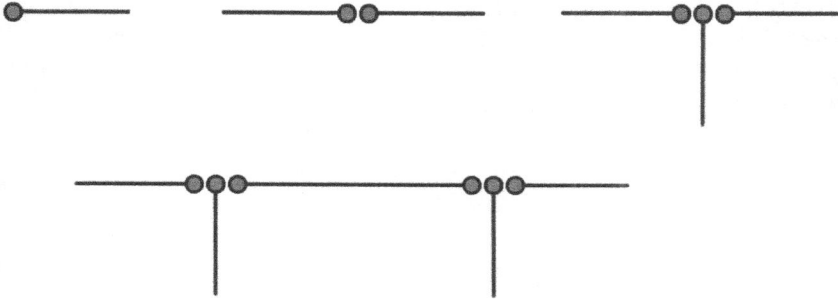

Figure 12: Three atomic relations.

Hooks are potentially occupied and closed, which means they potentially create relations in the actualization of potentiality: when closed, the hook is bonded creating a stable structure/relation; when open, it is ready to bond. Thus the name for such a hook becomes Potential Hook or POTHOOK (Fig. 13). Potential hooks are labeled according to the notation of Bisanz et al. (ISP 2019) as permanently closed hooks (ρ_{cl}), again open hooks (ρ_o) or inactive but activable pothooks (ρ). This notation permits the formalization of concrete cases of activation or deactivation of hooks. POTHOOK is schematized in the Beta Graphs as a pipe. This is mostly for the application to neuron activity but is not particularly relevant for the purposes of molecular biology, therefore I limit my usage of this tool only to hooks and not pipes.

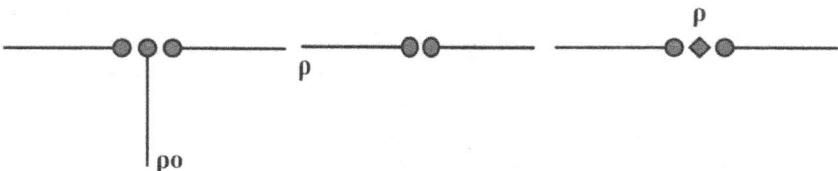

Figure 13: Pothooks

The branch is another important notion from the Betagraphic for the ISP. A branch is simply a divided stem, but the important note about the branch is that it cannot be generated from only stems, meaning that only a branch can be the generator for other branches. This recalls Peirce's NonReduction Theorem, to be addressed again in the following chapter. The potential stem or "P-STEM" is "a triad with two active common hooks, and one inactive pothook" (ISP 2019: 40). The P-STEM is "a virtual dyad, but under particular conditions, its inactive third pothook can become active" (ISP 2019: 40). This means that notions of activation and deactivation of hooks are introduced (see Fig. 14).

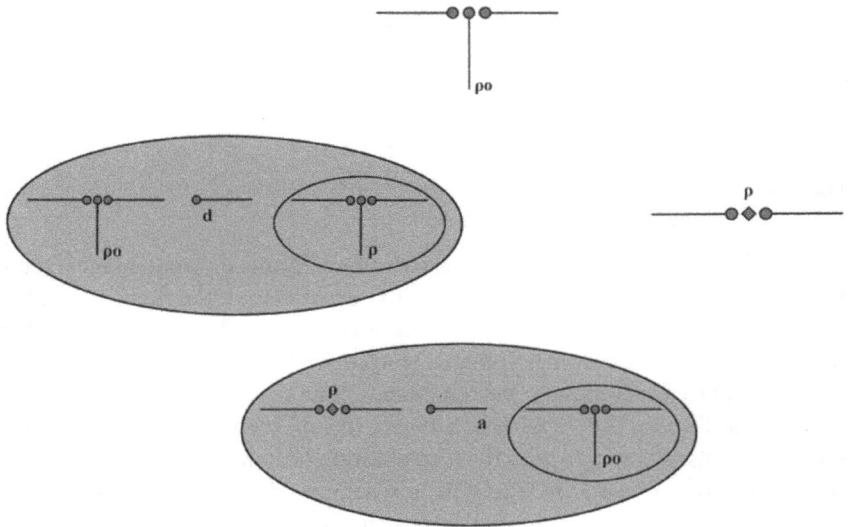

Figure 14: Activation and de-activation of pothooks.

In the figure, we can see in the first line a triad where one of the relates is an open potential hook. In the second line the open hook is deactivated by a deactivator labeled as "d". The circles represent the operation of implication: the outer circle implies the inner circle, which means in the case of deactivation of an open pothook, that the pothook becomes inactive but activable. In the very last line the process of activation is schematized: the inactive but activable pothook becomes activated by an activator labeled as "a".

Selected Examples

Betagraphic system of diagrammatic logical notation might bring new light into the understanding of certain biological phenomena under consideration. I will use Betagraphic as developed by ISP, but it is important to stress that this graphical notation is slightly different from Peirce's original Beta system, adopted by ISP to correspond with the current trends and discoveries in science. One of the differences is the addition of the pothooks as logical means for expressing potential relations, since these were identified as real in nature. Potential relations can be understood as non-existing reals (ISP 2009, 32). Phenomena which were not fully understood or exhaustively explained by current theories are actually describable diagrammatically by logical relations. The following examples are inspired by application to neurons by Bisanz et al. (ISP 2019) and are just another small step in the enterprise of Peirce's interdisciplinary scientific project[31]. As the authors of ISP 2019 applied the diagrammatic notation to resolve The Binding Problem in neuroscience with the supremacy of graphical representations of particularly complex relations in the neural system, I will use Betagraphic system (slightly modified from Peirce's original Beta) for the purpose of including potential relations to equally complex relations in molecular biology. Subsequent examples of formalization from molecular biology will be phenomena based on experimental research such as protein folding, splicing, epigenetic methylation on cytosine, ribosomal frameshifting and enzyme-substrate binding.

Gene Expression and Epigenetic Modifications

Gene expression is the best instance of the potential nature of biological organisms: the genetic information encoded in the texts written in the alphabet of nucleobases is only potentially transcribed, translated and therefore expressed in the form of concrete phenotypes.

> As a result of allowing for triadicity, one might find that genetic production of next generation basic-level neurons could reproduce genetic changes from the earlier survival experiences of one's ancestors (for example, food preferences, or taste in music or art). These in turn may manifest in a given individual's current thinking or behavior, even generations later. Interestingly, from a simple dyadic perspective, these characteristics or capabilities may appear "unexpectedly" and/or without apparent "cause". In reality, however, these subtle behavioral aspects could have evolved over time and have subsequently been coded as activatable gene sequences passed on from generation to

---

31 Work on Betagraphic has been developed in three volumes, in 2011, 2015 and 2019 by the Interdisciplinary Seminar on Peirce (ISP) authors.

generation. Such phenomena could be scientifically explained only if some neurons possess triadic relationship potentiality. (Bisanz et al. 2019: 54–55)

Potentiality related to gene expression depends on environmental context and external factors, but this is not the only potentiality going on at the DNA level. Besides genetic potentiality, we can talk about epigenetic potentiality, which is a concept even more context-dependent and even more potential[32]. As a matter of fact, every cell of an organism contains the same genetic information, so if it would depend only on the genetic information, logically all cells would be exactly the same. Besides the fact that genes might be expressed or not, genetic information can be ultimately modified by epigenetic marks. Epigenetic marking can take place at the DNA level, in form of methylation on cytosine or the adding of a methyl group to the cytosine which represents an additional alphabet to be added to the four nucleobase alphabet of the genetic script.

DNA methylation on the fifth position of cytosine (5mC) is a stable epigenetic mark that has important roles in mammalian development, differentiation and maintenance of cellular identity through the control of gene expression. (Kim and Costello 2017)

The word epigenetics is used because its object is valid and expressed information, but which is not encoded by the four famous nucleic bases. It has been experimentally proven that epigenetic marks might be inherited, but also can be reversible (Markoš and Švorcová 2019).

It has to be mentioned that by the notion epigenetics a lot of molecular processes are covered, not only cytosine methylation, but for instance also post-translational modification of histone and chromatin and informational processing mediated via various types of RNA molecules (RNA molecules might be the most important). Histone is a protein used for packing DNA into the nucleosome. Epigenetics studies many processes involving histone modification in the process of transcription, one of them being methylation. The consequences of methylation can be either positive or negative with respect to transcriptional activity, depending on the position of the modified residue within the histone tail. Some epigenetic modifications are associated with activation, whereas others are associated with repression, see Fig. 15 (Markoš and Švorcová 2009; Gibney and Nolan 2010).

---

32   There is a radical possibility for not differentiating between genetic and epigenetic, but according to the current trends in biology, we tend to understand genetic and epigenetic as qualitatively different categories.

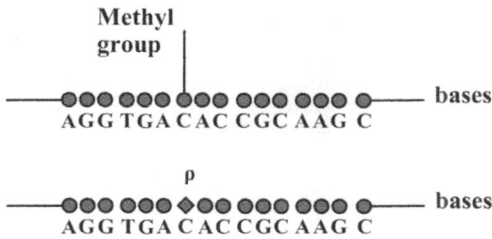

Figure 15: DNA cytosine methylation as a heritable epigenetic mark.

Splicing

In the previous chapter the process of splicing of the primary RNA transcript was mentioned. To recall this step in protein synthesis, we may repeat that after the transcription of the DNA string into the RNA string (primary transcript), the latter is ultimately modified, concretely unimportant parts are cut off, as a result only the coding part of the RNA transcript remains. The parts of the transcript that are cut off are called introns while the remaining parts are called exons. This process of cutting of unimportant parts of the primary transcript is called splicing. Spicing occurs thanks to special protein-RNA complex called spliceosome. In Betagraphic schematization we can label the spliceosome as an activator of splicing, as illustrated in Fig. 16.

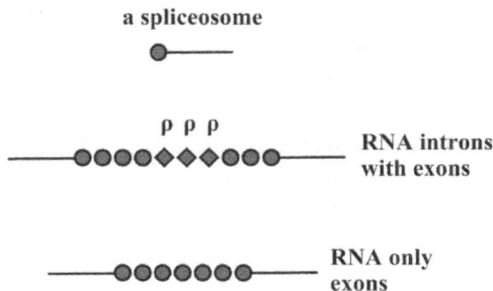

Figure 16: Activation of splicing by spliceosome.

Denaturation

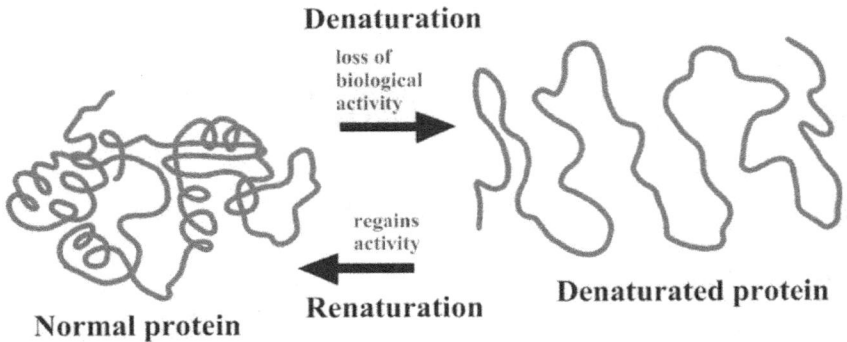

Figure 17: Denaturation of protein.

Protein Folding

Protein folding and its relatedness to the concept of potentiality was already discussed previously. The potential of the peptide chain to fold is somehow inherent in the structure of the amino acid string. Once the protein is folded, it can be unfolded; this might happen naturally by heat or with the aid of a solvent. The term used here is protein denaturation. When a protein is denatured, secondary and tertiary structures are altered, yet the bonds of the peptide chain – the bonds between amino acids – remain intact. When denatured, proteins can return to their folded (or "native") state again. This process is called renaturation (Fig. 17). A classic example of denatured protein is represented by boiling eggs. Egg whites contain the protein albumin. Fresh egg whites are transparent and liquid. Cooking egg whites turns them opaque; this is the result of denatured albumin fibres that interconnect between themselves, thus creating a solid opaque mass. Examples of Betagraphic schematizations of protein folding and denaturation are illustrated in Figs. 18 and 19.

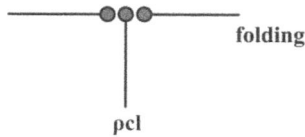

Figure 18: Protein folding in Betagraphic: when the peptide chain folds to its target conformation, the binding places- hooks for close and distant amino acids binding become closed.

Denaturation of protein

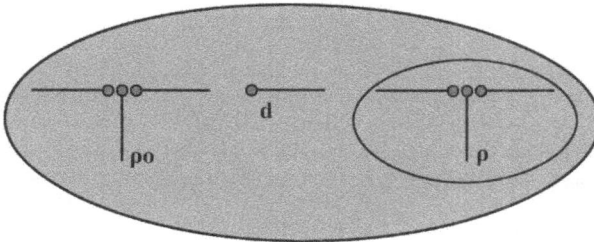

Figure 19: Re-opening of hooks in a protein denaturation. In the moment of a denaturation of a protein, with deactivators, hooks might be open again.

Ribosomal Frameshifting

Another molecular process I would like to schematize with Betagraphic is ribosomal frameshifting (Fig. 20). Frameshifting occurs when the ribosome, during its reading or translation of the RNA string, occasionally skips one base. The action of the ribosome is not accidental; it leads to a correction of the reading frame. It could be said that the ribosome disobeys the instruction to simply read base by base, and instead changes the reading frame by a goal-oriented activity. This action can be compared with a driver who disobeys the traffic rule to go when the light is green, when suddenly a child decides to cross the street and stops even if no red light nor stop sign is present. In this example, the driver decides to obey a different set of rules (not the traffic rules) in order to avoid an accident, similarly to ribosome avoiding an erroneous translation.

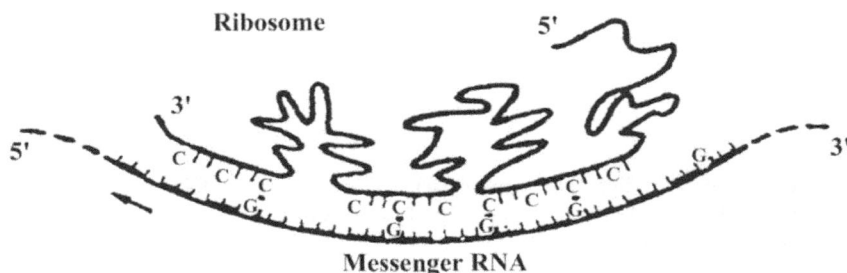

Scheme of the translation frame-monitoring mechanism.

Figure 20: Ribosome and the translation frame. From Trifonov 1987.

> Ribosomal frameshifting […] is a process where specific signals in the mRNA instruct
> the ribosome to change reading frame from the 0 to the −1 frame (movement 5′-wards)
> at a certain efficiency and to continue translation in the new frame. Frameshift signals
> are thus found within overlapping coding sequences. (Brierley, Dos Ramos 2006, 29)

Changing the reading frame from 0 to -1 means that the ribosome does not start reading from the first base but from the -1 base, which corresponds to reading by two and not three bases. Frameshifting was studied already in 1987 by Edward Trifonov, the founder of Israeli bioinformatics and the main proponent of the study field called DNA linguistics, today practiced by A. Bolshoy and his research group.

As is mentioned in the quote by Brierley and Dos Ramos, frameshifting is a process related to the overlapping of codes at the DNA level, which are cases when several messages are encoded by the same string, but through different reading frames. For more about the overlapping, see Bolshoy et al. (2004) or Bolshoy (2018).

We still do not know why and when exactly frameshifting comes into play, but probably this is connected to the recognition of the actual situation: it is recognized that the instruction "read bases by three" has to be violated by the influence of some particular conditions. We can call this the pragmatic level of genetic semeiosis. In this case, we do not have knowledge of what precisely might be the "activating" factor of the shifting step, therefore in terms of notation from ISP 2019, we can label this situation as "emergence from the inside" (2019: 44). What emerges is the relation between a base and another base, these being distant one from another, as a violation of the classical reading frame of the ribosome. The bases remaining unread, or "skipped", are annotated with inactive pothooks, since they are potentially active (meaning read by the ribosome and transcribed) and only in the actualization of frameshifting will they be potentially skipped (Fig. 21).

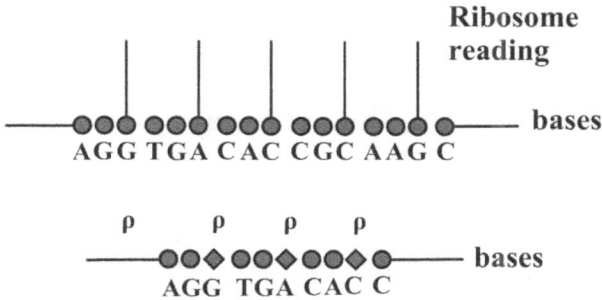

Figure 21: Ribosomal frameshifting with pothooks representing skipped bases, being technically inactive but potentially activable.

Enzymes

In the case of enzymes, the potentiality resides in the binding between the enzyme's active site and its substrate molecule (Fig. 22). This is the classic example of key-and-hole fitting between biological shapes.

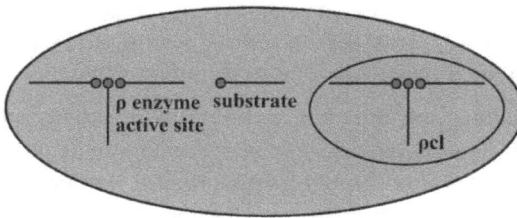

Figure 22: Enzyme interacts with the substrate: by folding into its native conformation, an enzyme acquires its characteristic shape to which a substrate fits perfectly. The binding between the enzyme and its substrate happens thanks to chemical bonds – these are represented by pothooks.

With some examples from the Betagraphic applied to molecular biology, we can see how Peirce's relational logic is perhaps better than linear symbolic notation for dealing with some contemporary biological questions. Epigenetic

potentialities and other triadic relations are illustrated by the Beta Graphs, conserving their triadic nature without violating the triadic relation by imposed linear order. There is no linear order, no hierarchy between the numbers one, two,
three. The order is logical or semeiotic and is from one to three to two (Pelkey
2012, see also Ketner 1989).

Continuity: Continuous Line of Identity

Potentiality is an important structural feature of molecular processes related to
protein biosynthesis and it is possible to diagrammatize it formally. Above, I proposed a small example of possible formalization of potential relations in biological structures in the variant of the Betagraphic's iconic notation by Bisanz et al.
(2019). Since the arrival of quantum mechanics, the principle of potentiality has
become a part of empirical and experimental science. The concept of potentiality
is an inevitable part of molecular biology and biology in general, given the fact
that it is a science studying living organisms evolvable and evolving (both in
terms of ontogeny and phylogeny) in time and space, meaning that the evolving
is by definition related to a momentarily unknown future. Potentiality is closely
related to continuity in Peirce's philosophy of synechism which is nothing but
thirdness:

> For although tychism does enter into it, it only enters as subsidiary to that which is
> really, as I regard it, the characteristic of my doctrine, namely, that I chiefly insist upon
> continuity, or Thirdness, and, in order to secure to Thirdness its really commanding
> function, I find it indispensable fully [to] recognize that it is a third, and that Firstness,
> or chance, and Secondness, or Brute reaction, are other elements, without the indepen-
> dence of which Thirdness would not have anything upon which to operate. Accordingly,
> I like to call my theory Synechism, because it rests on the study of continuity. (CP 6.202)

In this part I will focus on the term of continuity in various aspects of Peirce's
thinking in order to apply the principle of continuity to the protein folding process in the next part of this chapter. It's important to comprehend continuity in
a broad sense for understanding the overall achievement of Peirce. At the end of
the day, continuity permeates the various disciplines of Peirce's interest in independently one from another; it also guarantees a continual passing from one to
another in the spirit of a real interdisciplinary approach towards science.

Considering the very core of synechism as thirdness, we can understand it as
an imaginary interpretant connecting semeioses to other semeioses and developing new semeioses in nature. Living organisms are but representamen and
at the same time interpretants (Insc) of the immense continuity of nature. The
notion of synechism connected to biology and evolution of the cosmos is dealt

with minutely by Brier (2019). The theory of synechism is closely related to the tychastic and agapastic evolution, which was described in the second chapter of this book:

> Peirce goes beyond scientism through his triadic process non-dual view on reality as a semiotic field in that synechism can have no absolutes, such as mechanical Laws, atoms/ elementary particles or "Ding an sich"! (Brier 2019: 61)

In connection with some notions from Chapter 2 of this book, synechism might be understood as defined by Brier:

> The material, the habit, the signs and evolutionary forces are connected on a deep onto-logical level [...] Finally, he [Peirce] is a hylozoist, meaning that he thinks that life is also present within matter and that it evolves with it (Brier 2019: 64)

For a brilliant excursion into mathematical notions in Peirce see Chapters 3, 5 and 5 of F. Scott's book (Scott 2009). Scott also lists, besides continuity and synechism, the absolute and chance as Peirce's main mathematical conceptions. Continuity is for Peirce a baseline guiding his theories in metaphysics, semeiotic and of course in mathematics as well.

> By the mid-1880's Peirce recognized that a more satisfactory logical account of conti-nuity was needed. This would involve defining a certain kind of infinity which, in truth, required developing the logical doctrine of infinite multitude. (Scott 2009: 78)

In the appendix to his book from 2007 Stjernfelt discusses continuity in Peirce's writings in the most general way. Continuity, as the interdisciplinary permeable element in Peirce, is incarnated in mathematics as the notion of mathemat-ical continuum, in metaphysics as the theory of synechism, and in neurosci-ence (even though it is not possible to talk about neuroscience as such in the times of Peirce) as the famous and already mentioned Law of Mind. The Law of Mind (1892) is usually classified within Peirce's essays on evolution and it is also understood as such. Evolutionary Love is driven by a law, this law being built upon forces of attraction between objects and organisms in the cosmos, thus developing relations between each other and mediately, by the action of interpre-tants or thirdness since relations also exist across long distances in space as well as time. This is what Peirce calls evolution, Love, but it is first of all continuity. It is a continuity of space, time, mind, and matter. Minds are interconnected to each other and in this way ideas are shared by people – the more they are shared the less concrete and more general they become: this is the law of mind as formulated by Peirce. In a like manner, we could say, genes are shared by organ-isms. Genetic information is what the whole biosphere has in common – a high percentage of DNA is shared by all different species of plant, flies, mammals.

The basis of genetic information is spread all around the biosphere and is very general and less intense and less individualized in the way that it does not create particularities for specific species. Humans share, for example, 99 % of their DNA with chimpanzees and 92 % of their DNA with mice. Very little denotes the difference: in terms of DNA, the difference between humans and chimpanzees is only 1 %. The continuity in the genetic information shared among organisms is its crucial feature. Yet apparently it is not only the continuity of DNA that organisms share. We can also talk about the continuity of the cell membrane. Continuity in the law of mind was mostly ideated in the direction of explaining consciousness in particular, and the continuity of ideas:

> not only is consciousness continuous in a subjective sense, that is, considered as a subject or substance having the attribute of duration, but also, because it is immediate consciousness, its object is ipso facto continuous. (EP1: 315; 6.111)

Nonetheless, even if originally being Peirce's instrument to explain continuity of ideas and minds, it becomes the law of the whole cosmos, that is why it also makes part of Peirce's evolutionary theory: the law of mind is the driving force of Evolutionary Love. Consequently, we cannot reduce continuity to minds only.

Continuity exists also at the lower levels of living organisms. As an example I can mention the continuity of language, referring to Prodi's Copernican revolution in biology (see also Zengiaro 2022): there would be no language in human beings if there was no language in lower levels of life. The statement that a cell or a ribosome reads does not mean that it simulates the human act of reading in any kind of anthropomorphic and logocentric way, but rather that the human reading has origins in molecular interpretative processes (Prodi 1989: 25–26). Similar reasoning is nothing but a preliminary for the message of the NonReduction Theorem, that no triad can be generated from dyads or monads. Hence, if consciousness is triadic, it is not because it was generated from dyadic relations. It must be an extension of other forms of triadic relations we can find in nature. In words of F. Stjernfelt, "the continuum is a primitive phenomenon which may not be derived from simpler phenomena" (Stjernfelt 2008: 395).

My argument here is that, at a biological level, we can find triads and continuity already at the lowest life constituents: proteins. This continuity exists in the form of continuous lines, molecular chains that constitute basic life structures. Peirce was working on the mathematical conception of continuum mostly in the years around 1900, but even earlier than that we can find some interesting ideas contributing to the mathematical discussion on continuum in the last decades of the 19th century.

I will take the example of the geometrical form of continuity that is the continuous line. As Peirce himself states, the best way of treating the problem of continuity in philosophy should be led by geometry (NEM III: 101). The problematic part about the continuous line is its breaking, since the breaking of the line, which was until the moment of the breaking just an amorphous continuum without single points, leads actually and eventually to the creation of singular points, actualities, distinct identities. Actual existence does not occur unless there is something that interrupts the continuity, which is why, within the continuity, the principle of excluded middle is not violated: if actual existence does not exist, what only exist within the continuity is the potentiality, and potentiality means the coexistence of contradictory facts.

> Now if we are to accept the common sense idea of continuity (after correcting its vagueness and fixing it to mean something) we must either say that a continuous line contains no points or we must say that the principle of excluded middle does not hold of these points. The principle of excluded middle only applies to an individual (for it is not true that "Any man is wise" nor that "Any man is not wise"). But places, being mere possibles without actual existence, are not individuals. Hence a point or indivisible place really does not exist unless there actually is something there to mark it, which, if there is, interrupts the continuity. (CP 6.168)

Peirce explains his mathematical theory of continuum mostly as a disagreement about the famous Dedekind cut:

> Personally, I agree entirely with James, against Dedekind's view; and hold that there would be no actually existent points in an existent continuum, and that if a point were placed in a continuum it would constitute a breach of the continuity. Of course, there is a possible, or potential, point-place wherever a point might be placed; but that which only may be is necessarily thereby indefinite, and as such, and in so far, and in those respects, as it is such, it is not subject to the principle of contradiction, just as the negation of a may-be, which is of course a must-be, (I mean that if "S may be P" is untrue, then "S must be non-P" is true), in those respects in which it is such, is not subject to the principle of excluded middle. (CP 6.182)

If we imagine a continuum as a continuous line, Dedekind's comprehension of cutting a line consists in that the cutting point, the border between parts in relations to each other (two parts of a line), must always be assigned to only one of the two parts, so as to be able to be always reduced to a determined individual entity. For Peirce, the cutting point does not belong to either of the two parts, or rather, belongs to both of them, as represented in Fig. 23 (see also Paolucci 2004: 125).

A	P	D A	B	C	D	

A	P	D A	P	D	A	P	D A	P	D

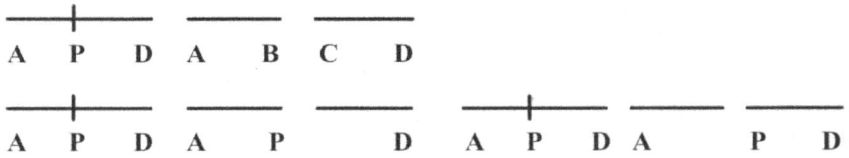

Figure 23: Cutting a line by Peirce and by Dedekind: the Peircean cut is represented by the upper line (AB, CD) and Dedekind's cut is represented by the lower line (AP, D or A, PD). See also Ketner and Putnam 1992, 39–41.

> In Lecture Three Peirce said a number of things which sound completely bizarre to today's mathematical sensibility. Then he challenged anyone who might think he is crazy to show that he had actually contradicted himself. What is worse, it looks very easy to show that he has indeed contradicted himself. (Ketner and Putnam 1992: 41)

The authors are referring to the Cambridge Conferences Lectures from 1898. In the next pages the authors give examples and explanations of such potential contradictions related to the alternative to the Dedekind cut and proposed solutions as well:

> Notice, by the way, that in his formulation of the paradox Peirce said that the points have been at one point, and that the symbolization "=" is not a natural one for being at one point, although it does sound natural for the other locution Peirce used, apparently interchangeably, of having been one point. Another problem, of course, is that Peirce spoke as if things could be identical at one time but not at another, yet in the Logic of Relatives, as in present-day predicate Calculus, identity is treated as a tenseless relation. But if "being at one point" is not an identity relation, even for points, then this problem disappears. (Ketner and Putnam 1992: 43)

The problem of the identity of a single point of a line is proposed to be resolved by not considering "being at one point" as identity relation. It might be interesting at this point to evoke the notion of mereology, discussed by Stjernfelt:

> Thus, when Peirce claims that the infinite series of points "are" one single point, then this is no identity claim, but rather a mereological use of "to be" so that it should be read as saying that the point "consists of" of many points. (Stjernfelt 2007: 407)

I will just recapitulate that Continuity is necessarily related to potentiality which is hidden in every living being. It is hidden in the unexpressed genes contained by the genome of one's organism, in the evolving of one's organism in a given way depending on external conditions, in the unpredictable future: the law of mind has different rules than physical laws – every living being is indefinite in its potentialities. Potentiality exists also in geometry, namely in the dividing of the line. Ketner and Putnam mention potentiality in the Aristotelian sense: when dividing the continuity, the cut point potentially belongs to one or another part of the divided continuous line. Additionally, the paradox of cutting a geometrical line leads to further metaphysical problems, such as the question of the identity and singularity of the existence of points:

> Why can't the line be divided by putting the monad of P in the right half of the division, together with every point which lies to the right of P by some finite amount, and putting all the other points (the points which lie to the left of P by some finite amount) in the left half of the division? It would seem that this would divide the line cleanly into two pieces in such a way that the left hand of the division would contain no greatest "point", that is, it would contain no greatest monad. This would violate both the Aristotelian intuition and that much of it to which, as we have seen, Peirce himself is definitely committed – namely, that if we divide a line or a line interval and then separate the two segments, then the left half of the division will have a right-hand endpoint. We believe we know the answer to this question, and not surprisingly it takes us deeper into Peirce's meta-physics than anything we have talked about so far, because what we have said so far does not really appeal to Peirce's metaphysics. It appeals only to his belief in the very large cardinality of the points on the line. In particular, we have not so far interpreted Peirce's mysterious idea that when a multitude becomes that large, then its individuals lose their distinct identities. (Ketner and Putnam 1992: 49)

In the words of Peirce,

> But the line is a mere conception. It is nothing but that which it can show, and therefore it follows that if there were no discontinuity there would be no distinct point there, – that is, no point absolutely distinct in its being from all others. (Peirce 1992: 160)

A line is a multitude of points, but a multitude so large that single points have no distinct identities. In Lecture Eight Peirce concludes that this kind of aggregate of individuals results in the conception of potentiality. Thus it is a "collection of *possibilia*" (Ketner and Putnam 1992: 51) which are not determinate objects for Peirce, but are rather relations. Thus, the continuity is defined in terms of relations, and it is topology which brings us back to semeiotic as relational logic.

Folding the Continuum

Peirce's version of the geometrical understanding of the continuous line pre-supposes that every cut of the line creates cutting points, but these remain undecided: the identity of the cutting point is assigned to neither of the created segments of the line, or rather, it belongs to both of them. This creates a logical paradox of the violation of the law of the excluded middle and might be a diffi-cult concept for classical mathematical views on the continuous line, as already mentioned. An elaboration of Peirce's continuity has been proposed by Paolucci (2004), who considers the concept of *folding* rather than cutting the continuous line. In fact, cutting represents the moment of breaking down the mathematical continuum. A cut is very radical in the sense that it necessarily constitutes a violation of the excluded middle. But when it comes to *folding*, the situation is different. Folding leads to breaking down the continuum without losing its continuity. The semeiotic concept of folding of the continuum proposed by Paolucci is a means of breaking the continuum without creating cutting points, thus somehow avoiding the mathematical paradoxes or potential inconsistencies.

In fact, what Dedekind's model represents is simply a cut. Contrary to this, Peirce's model of "cut" can rather be replaced by the notion of fold, because only the folding of a line or of a continuous surface breaks down the continuum and at the same time conserves its continuity. Furthermore, the points of the folded line (contrary to the cut line) are undecided; they exist in the mode of the conditional, of may be. In fact, they are not points in the sense that they are not single, individually existing identities. G. Deleuze was also fascinated by the notion of the fold, to which he dedicated the essay *The Fold. Leibniz and the Baroque* (Deleuze 1988). Deleuze maintains that the fold, not the point, is in fact the smallest element of matter, his observation being similar to Peirce's critique of Dedekind's cut:

> The unit of matter, the smallest element of the labyrinth, is the fold, not the point which is never a part, but a simple extremity of the line. That is why parts of matter are masses or aggregates, as a correlative to elastic compressive force. Unfolding is thus not the contrary of folding, but follows the fold up to the following fold. (Deleuze 1988: 6)

Now, I would like to continue with my thought experiment and to proceed to the application of the theory of a folded continuum to the "organic strata". As was previously proposed, a peptide chain could be, for the sake of the semeiotic interpretation of biological phenomena, comprehended as a continuous line, unarticulated mass, or undecided possibility of potential folding. The peptide chain is, in our experiment, a "body without organs" unless it is folded. Once it folds, the continuum is broken by the action of folding and dyadicity emerges;

folds are points of frontier between undecided potential triads (see the formalisation in the previous chapter) and decided dyadic relations between concrete amino acids in a chain. Possibility turns into actuality: a protein has its shape and it cannot have had another shape, at least unless it is denatured and refolded again or unless it interacts with another molecule, a substrate for example. Consequently, we can talk about the dyadic relations of expression by which content is encoded. We can talk about amino acids as discrete units of organic expression which encode the organic content represented by a protein's function, or by the protein's shape, because, let's say, a protein structure and its function are coextensive notions. If applied to molecular biology, "the point consisting of many points" is every single amino acid which by itself is constituted by other molecules and creates higher units consisting of many amino acids: the peptide chain.

Deleuze compares the process of folding the continuum to origami, the art of paper folding. Folding, unlike cutting, preserves the continuity of the paper sheet. Take an origami crane, unfold it carefully and flatten it out; it will become an unarticulated square sheet of paper. Take a protein and put it in an organic solvent, or heat it; it will dissolve the bonds responsible for folding, it will become a line again. Protein denaturation and consequent renaturation is similar to an unfolded sheet of paper which one can fold again to regain the initial paper crane.

Protein folding is a perfect example of a folded continuum which, even when creating binary relations (between one and another part of a folding point), preserves its triadic continuity, since it does not break down peptide bonds between amino acids.

A protein is like a continuum, constituted by strong chemical bonds which never dissolve and bonds creating folds, potentially unfolding and refolding again. Moreover, only when the continuum is folded is discontinuity generated; only at this point can we identify amino acids as "single points", singular individualities responsible for the folds. Even if chemical bonds (peptide bonds) exist between every neighboring amino acid, in the complex of structured proteins they are irrelevant – they are but the brute material. Chemical bonds between distant amino acids creating folds are on the contrary important for deciding the protein's final shape and function.

With aid of the notion of folding, Paolucci explains how the coexistence of continuity and discontinuity is possible. The idea of folding a continuum is that when the continuum is folded, it never breaks its continuity, it is just folded. "From continuum and in continuum, without ever interrupting continuity, it is possible to create discontinuity effects that give rise to opposition relations" (Paolucci 2004: 135).

Going Beyond: Gamma Graphs

Synechism is the structure of phaneron, but also, as has been demonstrated, it is the structure of proteins, the smallest functional parts of living organisms. As Prodi extended molecular language to human language, we can see the extension of the structure of the proteids in the structure of phaneron. The structure of our reasoning follows the same elementary pattern of proteic structures: it is governed by the folded continuum, which never breaks its continuity but creates the "differences", singularities thanks to folds. This kind of structure can be represented in the form of graphs. Recursivity of the elementary graphs expands from the molecular level all the way to reasoning. According to Peirce, diagrams help us to acquire a different way of thinking, a clearer form of thinking.

> A person who has learned to think in beta graphs has ideas of the utmost clearness and precision which it is practically impossible to communicate to the mind of a person who has not that advantage. (MS 467: 4)

While some kinds of reasoning are so simple and automated that we don't realize we are performing diagrams, more complex reasoning requires awareness and visualization of the diagrams. This is why Peirce decided to provide and design the graphical-visual models for diagrammatic reasoning, his Existential Graphs.

> Where the syllogism was a very simple one, reasoners were able to jump from the premises to the conclusion without realizing that, in fact, they had resorted to a mental diagram to perform the reasoning. To show how necessary conclusions are actually drawn, Peirce used a more difficult syllogism in which the conclusions were not so evident, and which clearly required observation of a diagram. (Scott 2009: 108)

The Beta Graphs are potential graphical tools allowing for representations of very complex relations. They still do not perfectly cover all the complex and manifold relations existing in synechism and the phaneron. To represent the relations of potentiality graphically, fully preserving continuity, recursivity and randomness was a task Peirce didn't have time enough in his lifetime to complete. Here, I am referring specifically to Peirce's so-called Gamma Graphs. The incomplete Gamma Graphs (GGs) were supposed to represent every aspect of synechism. Probably the most striking innovation when compared to the Alpha or Beta Graphs is that the GGs were designed specifically to represent relations of potentiality. As was explained before, this aspect is missing in Peirce's original Beta Graphs, and it was only additionally presented as an experiment by Bisanz, Ketner and their colleagues from the Institute of Studies in Pragmaticism, in Betagraphic, in the form of potential hooks (pothooks). Peirce himself did not include such features in the Beta Graphs, but hypothesized them for the GGs. As will be shown, the GGs are in their hypothetical existence even more suitable for diagrammatic description of the living, including tools for diagrammatization of epigenetics and real protein possibilities.

The only source of insight into the GGs is MS 467, containing Lecture Four of *the Lowell Lectures* by Peirce from 1903, recently published in the volume *Lowell Lectures of 1903 by Charles S. Peirce* edited by Ketner (Ketner 2024, 85–102).

The lecture was entitled *Existential Graphs, Gamma Part*. It starts with a general introduction to EGs, describing in detail from the first pages the particularities of the Alpha and Beta Graphs and starting the Gamma part at page 18 of the manuscript, with a foreword about the general recursivity of graphical thinking:

> How many people fancy that they know a language very well when they can think in it. If they pursue the study, they will afterward turn back and see that that was merely the time at which their real knowledge of the language was making its first beginning! It is precisely so in graphs. When you have learned to think in them easily without translation then you are ready to begin the real study of them. (MS 467: 18)

Here Peirce draws an analogy between learning a second language and learning a graphical logical notation. Not only is there recursivity in reasoning, in that every reasoning is diagrammatical as it follows the diagrammatical structure of the phaneron, but also the notational graphical EGs are recursive in that we can only can learn them if we are already reasoning in them. But how can we start reasoning in GGs without understanding the system of its graphical notation? This paradox becomes even greater when we realize that Peirce didn't really provide a detailed description of the graphical notation for the GGs. The reasons are multiple: first, he didn't have time enough to accomplish the project during his lifetime; secondly, it is questionable whether the nature of the GGs allows for any kind of (purely) visual schematizations as is the case of the Beta and Alpha Graphs. What is particular for the GGs in terms of their visual representations, compared to the Alpha or Beta Graphs, is the complexity of the relations they represent and the impossibility of reducing these relations to two- or even three-dimensional space:

> I ask you to imagine all true propositions to have been formulated, and since facts blend into one another, it can only be in a continuum that we can conceive this to be done. This continuum must clearly have more dimensions than a surface or even than a solid, and we will suppose it to be plastic, so that it can be deformed in all sorts of ways without the continuity and connection of parts being ever ruptured. (MS 467: 24–25)

Here Peirce talks about a topology of space having more than three dimensions. Even this kind of space, so hard to imagine, has topological properties; that is, it is a topological space that can be deformed without being ruptured, or without destroying the relations of the points constructing the space. Here again Peirce talks about topology without explicitly naming it. Peirce then gives as a simplified example of such a continuum: a photograph is a deformation of the plastic, dynamic object of that photograph. Peirce's semeiosis model can be very easily explained in this manner, with topological geometry. The relation between the dynamic object and the representamen is a topological relation. Another Peirce example is more classical, a map (MS 467: 28–33). He mentions the map even in the mathematical essays on topology (the problem of coloring a map, see fnew1–4). In both the cases of a photograph and a map we are representing a topological relation of points determining a space being collapsed into another space without rupture of relations between the points constituting the original space: for instance, in the relations between points determining the physical object photographed, the photograph itself is characterized by the very same relations, with the difference being that they are collapsed from three-dimensional space into two-dimensional space. When representing the relations of possibilities in a more than three-dimensional space in the GGs, we have to imagine an analogical situation of the collapsing of spaces one into another without the rupture of relations. In this case, the two spaces represent actual existence (a photograph of) and the real possibilities (what is photographed):

> But all the qualities any one [...] can think of are certainly innumerable, and all that might be thought of exceed, I am convinced, all multitude whatsoever. For they are mere logical possibilities, and possibilities are general, and no multitude can exhaust the narrowest kind of a general. Nevertheless, within limitations which include most ordinary purposes qualities may be treated as individuals. At any rate, however, they form an entirely different universe from the universe of existence. It is a universe of logical possibility. As we have seen, although the universe of existential fact can only be conceived as mapped upon a surface by each point of the surface representing a vast expanse of fact, yet we can conceive the facts are sufficiently separated upon the maps for all our purposes, and in the same sense the entire universe of logical possibilities might be conceived to be mapped upon a surface. Nevertheless, in order to represent to our minds, the relation between the universe of possibilities and the universe of actual existent facts, if we are going to think of the latter as a surface, we must think of the former as three-dimensional space in which any surface would represent all the facts that might exist in one existential universe. In endeavoring to begin the construction of graphs, what I had to do was to select from the enormous mass of ideas then suggested a small number convenient to use more than one actual sheet at one time, but it seemed that various different kinds of cuts would be wanted. (MS 467: 38–42)

Peirce then introduces the "broken cut", typical for GGs. It does not, contrary to the classical cut from the Alpha and Beta Graphs, represent a negation, but something like non-negation, or *a possibility of* negation, or *undecided* negation. The "broken cut" is a graphical representation of real possibilities: it is a negation which exists in potentiality. Indeed, relation (any kind of relation) itself is but a mere logical possibility (MS 467: 84). Thus, semeiotic as a relational science is all the time dealing with logical possibilities (also relations). Peirce's GGs are a viable graphical representation of real possibilities, that will be important for any extended semeiotic research.

We conclude that for Peirce the paradox of cutting a line extends beyond geometry and is anchored in metaphysics. Metaphysics and semeiotic as the methodological tool for Peirce represented an ongoing cutting edge of theoretical science across all sciences. In the words of Bisanz, the task semeiotic is "exploring new ways to enhance scientific research in humanities and consequently aiding to bridge the gap between Humanities and Natural Sciences" (Bisanz 2019: 1).

The singularity of points on a continuous line is an example of real possibilities. Single points exist potentially; they are real possibilities. They can be represented in the GGs in terms of broken cuts, but we do not know whether they exist as singularities or not. The multitude of points constructing a line becomes so great that we cannot distinguish single points anymore. But the unanswered question remains about *when* we know that the multitude is so great. *Who* decides when single points lose their particular identity? This is a semeiotic question to be answered with the help of the Gamma Graphs. In the work of Eco, and inspired by Deleuze, we find the distinction between the *molecular* and the *molar,* where the molecular represents the distinction between singular points, which are irrelevant for the interpretive process. For example, I recognize a table as a token of a certain type thanks to its shape and some characteristic properties (it has four legs and a certain dimensions and is made from a solid material), but not thanks to the chemical composition of the molecules of the wood it is constructed from. For Eco, semeiosis is not to be separated from interpretive process and *interpreter*. Thus, the interpretive subject is somehow necessary in distinguishing molar from molecular. It is also the case, it seems, in defining the multitudes great enough for single points to stop having their particular existence. I apply deliberately the logic of the Gamma Graphs to Peirce's mathematical essays in order to explain them from the metaphysical perspective of the mathematical notion of continuum.

There is indeed an allusion to the subjectivity present in MS 467. At page 60, Peirce says a "whale is not a fish", but this claim is only possible from the perspective of a third-person observer. An epistemic cut must be made between the observed object and the observer. This is a revolutionary experiment by Peirce. Introducing subjectivity and the first-person perspective to logic and to the EGs is surely an achievement of particular importance. It is true that in MS 467 the allusion to the first-person perspective is only implicit, but in my understanding it is crucial and it also explains the need for more than three dimensions for graphical representation of the GGs: the first-person perspective requires one more dimension to be added to the dimensions of the already collapsed space of the representamen (photograph/map). I argue that the addition of the dimension of the first-person perspective is also essential in understanding logic in continuity with semeiotic and living (growth): the epistemic cut between the observer and the observed object implies the very definition of life. As proposed by H. Pattee (2008), the epistemic cut is defined in relation to the genetic code. They encode something which is different from the code-makers; therefore, at the moment of the appearance of the genetic code, the epistemic cut was one of the preconditions of self-replication based on coding. The epistemic cut lies between the observer-subject and the observed objects. The epistemic cut between subject and object as a precondition of life was recently rediscovered and further commented upon Gazzaniga in the context of neuroscience (Gazzaniga 2018). Based on Von Neumann (1966) and Pattee (2008), life is a subjective phenomenon in that code usage necessarily requires a separation of first person from the third person. We can call the first-person perspective also interpreter's perspective.

In this way, subjectivity (as one of the characteristic features of the GGs and the epistemic cut) might represent a problem for mathematics and logic and their definition as "objective" sciences. Yet, when we understand mathematics as a predisposition for semeiotic, and in particular in the unfinished patterns of the GGs, subjectivity must be included in the metaphysical aspects of the geometrical paradox of the cutting of a continuous line. In the thought experiment I presented, the relation between metaphysics, subjective interpretation, and geometry is clear. Single points do exist in the peptide chain (single amino acids); this is an experimental fact we can easily demonstrate in any laboratory, but they exist at the molecular level from the perspective of protein biosynthesis. The single points then appear when the peptide chain is folded, only as a result of the interpretive process of the organism, creating its own self-reproduction and protein production as interpretation of both the virtual inherited script and the current environmental-contextual conditions (genetic and epigenetic). Protein synthesis is an example of the epistemic cut at the molecular level.

It should can be stressed that diagrams are not necessarily visual, nor are they necessarily two- or three-dimensional. In the GGs the visual is somehow not enough. Another kind of representation is needed which Peirce had no time to properly describe, for the nature of Gamma Graphs to be comprehend beyond the linear continuity of space and time. This leads me to yet another very special feature of the GGs. Besides the representation of potentialities and the subjective condition, the GGs touch a peculiar question of time.

It has to be said that when understanding diagrams in continuity with semeiotic and life phenomena, the notion of time is crucial. Growth is possible only in time, and we cannot treat growth as a tenseless relation. Therefore, the time aspect of the GGs requires further research. The symbolic nature of the living (organic codes as symbols which grow) is oriented towards the future. In topological representations of such relations, no paradox arises about the possible rupture of time when one space collapses into another, because of Peirce's very special opinion on the time-space relationship. In correspondence with Lady Welby, in the letter from January 7, 1905 (SS: 45–50), Peirce expresses skepticism about time and space continuity. He argues that time and space are related in our minds only because of language: the limits of verbal language restrict us to comprehend space in temporal terms, but it is not necessarily attached to the *real* relation of time and space.

Peirce concludes his MS 467 on the Gamma Graphs with a statement about recursivity, that we can only reason about graphs in graphs. Recursivity is only one of the particular features of the GGs, which are distinguishable also because of their time and subjectivity aspects. As a consequence, there is no direct instruction by Peirce on how to design or draw such GGs. Due to the impossibility of reducing the complex relations into two- or three-dimensional space, the graphical representations of the GGs are open to future research and investigation in the fields of logic and semeiotic; yet already now we can state that the GGs are the most suitable system for the formal description of living phenomena, because they provide tools for describing real possibilities.

Chapter 6 There Is No Form Without Substance: A Linguistic Analogy

The Language Metaphor of Life

Peirce understood life as defined both in terms of forms-diagrams and in terms of growth. The link between the two, the link connecting organic forms hidden in seeds and germs and the tendency to growth of living matter, remains blurred. Indeed, only with the deciphering of the genetic code are we finally able to achieve the idea Peirce couldn't foresee, to determine the link. Yet he provided us with the clue to define this link in terms of his semeiotic, with symbolic semeioses in particular. Symbols as specific types of semeioses defined by the quality of thirdness and the conventional relation (a habit-law) between representamen and object have been introduced with great glory to the scientific community and embraced since the sixties as the big answer to the mystery of life. The link between the potential of growth in every seed and the developed organic form is mediated by a repository, an enormously long text written in an alphabet composed of only four symbols. Peirce's famous "symbols grow" (CP 2.302) suddenly acquires a whole new meaning. The so-called language metaphor of life was born. And as a matter of fact, it was introduced by a linguist from the Prague linguistic circle, Roman Jakobson (1971). The metaphor had great success and immediately became a part of every genetic textbook. This metaphor was born in the nascent paradigm of generative grammar in linguistics based on the notions of information and code and it was related to the central dogma of molecular biology. But we are now living in a new era in biology; epigenetics questions the language metaphor of life. In parallel to the emergence of epigenetics and development in biology in the last few decades, we are experiencing a similar paradigm shift in linguistics: the very definition of language is being questioned in the same manner as the definition of the gene is being questioned in biology. In this light, the language metaphor of life also must be reconsidered. Starting with the question *what is language*, we can begin to redefine the language metaphor of life in the context of the 21st century. What about the blurred distinction between written and oral language with the arrival of technology and digital communication? Is written text a reduction of spoken language or is it vice versa? It is by no accident that the pragmatic turn in linguistics is accompanied by the epigenetic turn in genetics (see more on this topic in Lacková and Bolshoy 2021). How does the ambiguous and changing definition of language correspond to the situation in epigenetics and developmental studies? I answer some of these questions in the following excursion into linguistics.

Form and Substance, Phaneron and Proteins[33]

The recursivity of diagrams in the phaneron, the possibility to reason about semeioses only through semeioses and to comprehend diagrams only through diagrams – as summarized in MS 467 – is not just a particular feature of the Gamma Graphs, but also of Peirce's semeiotic in general. We can say that Peirce's account of semeiotic *is* formal: the Existential Graphs are but formalizations defining semeiotic relations in nature and in culture. But when dealing with forms, it might be interesting to have a look also at substances, to understand what was the relation of form to substance for Peirce. To do so, we have to go back to the manuscript titled What is Law of Nature? (MS 870)

A topic not often discussed in Peirce studies is the relation between form and substance. Without doubt, the motivation for this neglect originates in the tendency to draw a boundary line separating Peirce and European structuralism, or what is sometimes referred to as "semiology". The form-substance distinction was crucial for the structuralist approach, mainly for the theory of Louis Hjelmslev. Many structuralist thinkers relied upon Peirce. For example, Roman Jakobson directly refers to Pierce's works. Peirce himself also used the terms of form and substance and was interested in the relation between them, especially in his essays on evolution and growth in relation to the platonic forms and related aspects of organic development.

In biology the distinction between form and substance is essential. That is why the Platonic approach, often associated with formalism in biology, is so present and so often criticized (Markoš and Švorcová 2019). The platonic approach is nevertheless not the only possible way to address formalism in biology. We should distinguish the understanding of forms as predetermined (Platonic) from the more modest statement that there *are* forms in organic nature. In other words, the reality of forms in biology is undeniable: all the current research in proteomics and other biological disciplines is based upon this presupposition. The relational-topological nature of proteins is a perfect example of the importance of the form-substance question. Artificially produced proteins only support the argument that proteins are but forms, structures, where material realization is

---

33   This chapter has been partially published as Lacková, Ľudmila (2023). Structural semiology, Peirce, and biolinguistics. *Semiotica* 2023 (253): 1–21.

secondary. Biological forms have already been discussed in the context of real possibilities in nature in this book. The recursivity of diagrams might be also a way to address the continuity of forms in biology as a tool to help us understand the constitutional building principle of our mind, of the phaneron. Our thinking in diagrams might be continuous with diagrammatic forms at the molecular level of our cells, but where do the forms-diagrams come from? Like Kant's schemata, visual diagrams are substantiations of the "real", transcendental diagrams. Proteins as genetic diagrams are substantiations of "real molecular" diagrams existing before their embodiment in peptide structures; we need not go so far as to say that the "real diagrams" are products of Platonic idealism, but where can these organic diagrams be found? Peirce gives us the answer.

The elementary chemistry of all matter is diagrammatical: in the chemical structures of an atom, its chemical valency is a diagram which is in itself a proposition *John gives John to John*. In the elementary chemical structure, the difference between substance and form is blurred. Substance becomes form and form becomes substance, and they coexist in a singularity, in the potentiality of the matter-form to become a form for another matter or matter for another form. The "schemata" at the molecular level are substantial not in the sense that they are substantiated by the peptide matter (in this case they are already embodied diagrams) but are substantiated already in their chemical potentiality. In the case of an atom of ammonium, it becomes quite difficult to decide whether we are discussing form or substance. For Peirce, I would like to argue, form has precedence over substance and his highly developed visual diagrammatic logical system is a proof of this claim. In his doctrine of semeiosis, Peirce does not use the distinction form-substance (but he does use it in his writings on growth and evolution). Maybe it was not because he was disinterested in this kind of binarism, but that it was implicit in the very definition of semeiosis that its nature is structural, diagrammatical, and thus *formal*. A semeiosis has no other definition than the one through the mutual relations between representamen, object and interpretant. This definition is a structural one. By its very definition, Peirce's semeiosis is a structure taking place thanks to connections between relates in a triadic relation, no matter the substantial realization of the relates. Peirce was an excellent structuralist: his direct mentions of the dichotomy of substance-form are testimony of this claim. Structural thinking is also present in Pierce's work on the phaneron, or the constitution of our mind as relational, or diagrammatic. In MS 1334 (published in NEM IV) we find arguments for the precedence of form over matter in the structural-diagrammatical constitution of the phaneron, where it is thanks to form that we are able to reason:

> Form is something that the mind can "take in", assimilate, and comprehend; while Matter is always foreign to it, and though recognizable, is incomprehensible. The reason of this, again, is plain enough: Matter is that by virtue of which an object gains Existence, a fact known only by an Index, which is connected with the object only by virtue of brute force; while Form, being that by which the object is such as it is, is comprehensible. It follows that, assuming, that there are any indecomposable constituents of the Phaneron, since each of these has a definite Valency, or number of Pegs to its Graph-Instance, this is the only Form, or, at any rate, the only intelligible Form, the Element of the Phaneron can have, the Classification of Elements of the Phaneron must, in the first place, be classified according to their Valency, just as are the chemical elements. (NEM IV, 322)

In another part of the same manuscript, Peirce claims that "Form, in the sense of structure, is of far higher significance than the Material." When it comes to the formal classification of the elements of the Phaneron, we have two ways of classification of the decomposable elements of the Phaneron:

> one is as division according to the Form or Structure of the elements, the other according to their Matter [...] If, then, there be any formal division of elements of the Phaneron, there must be a division according to valency. (NEM IV, 322)

Here Peirce clearly states that the only way of classifying elements of the Phaneron is through valency. Peirce's valency theory was described earlier in connection with the logic of relatives. As mentioned in the previous chapter, the phaneron and proteins share, according to Peirce, the same structure constituted by valency. This is another example of Peirce's relational thinking: not only does form have precedence over matter, it is also *universal*: it is shared among minds and the smallest molecules of our bodies.

The acceptance of *forms* in nature has some negative connotations in contemporary science, mainly because of the stigma of Platonic idealism. See the discussion about Rádl by Markoš and Švorcová in the Chapter 2 of this book.

Revolution of Substance: Reversing the Order

To avoid vitalistic or creationist theories and retain some semblance of formalism, the only way to approach forms in biology is by reversing the order between form and substance. A paradox lies in that Peirce was a perfect example of a structuralist even though many scholars claim otherwise. On the other hand, some structuralist thinkers propose to reverse the order between form and substance. The topic is raised in the dissertation thesis by T. Bennett (2021) for example, who points out that this reversal is called by Greimas and Courtés "reciprocal presupposition" (146). Bennett uses the term *retroactivity* to describe the reversal. I argue that the reversal can also already be found already in Hjelmslev himself, upon whose

linguistic theory Bennett's notion of retroactivity relies. It was Hjelmslev to propose the terms of form and substance in semiology, and he was actually not that
strict about the hierarchy of the order as his later followers. Hjemlev proposed
a sign model inspired by the linguistic sign by F. de Saussure, yet expanding the
two components of signifier and signified to four components: form, substance,
content and expression. Hjelmslev emphasized the hierarchy between the two
couples of terms by talking about *form of* the content, *form of* the expression, *substance of* content and *substance of* expression and not vice versa. Thus, content and
expression are terms of higher order than the terms of form and substance. Even
if Hjelmslev's expression and content are not reducible to signifier and signified,
we can leave this couple aside and focus on the form and substance notions. It
is generally assumed and Hjelsmlev clearly stated in *Prolegomena* that there also
is a hierarchy between form and substance, in that form determines substance.
Yet in other writings by Hjelmslev the hierarchy between form and substance is
blurred, I will get to this particularity in a while. If we are to complete the analogy between Hjelmslev's model and the original model by Saussure, the form-
substance dichotomy can be to some extent compared to Saussure's dichotomy
of *langue* and *parole*. While form is more or less what Saussure meant by the
"langue", when it comes to substance, things get more complicated.

It is important to note that substance in Hjelmslev is not interchangeable with
the empirical matter of the referent, but is rather something like the "outside of
the inside" (Bennett 2021: 15, 48). The substance of the expression is easily understood: it is a matter of phonology as opposed to phonetics, not the really acoustic
or articulatory characteristics of sounds; but when it comes to the substance of
content, it is somehow more difficult to imagine what Hjelmslev meant by this
component of his model. Very simplified, we can assume that the substance of
content is *semantics*. Hjelmslev in his *La stratification du langage* defines the substance of content with an example of interlingual translation. When defining the
substance of content of a language, we need to relate language with other social
institutions or cultural artefacts. Not only do abstract contents have different
substances in different languages, but also physical natural objects have different
substances of content – in other words, they have different semantic substances.
He lists many examples of natural phenomena with one and the same referent
but different semantic substances across interlingual translations. He speaks for
instance about the Russian word "slon", meaning elephant. Elephant as semantic content differs significantly for speakers of African languages or speakers
from India, having many cultural or spiritual associations, while for European
countries the word "elephant" is mostly limited to associations from zoological gardens, circuses, etc. In relation to this observation, Hjelmslev states that

"substance is subdivided differently in different languages by form." In this claim, the important two words are *by form*. So to speak, there are two observations resulting from this quote: Firstly, substance is not empirical matter and is not the referent. Secondly, substance is divided *by form*. I will get to back to the second point later, but for now we can say that it is proof that Hjelmslev was not logocentric.

According to Bennett,

> Logocentrism is rather the belief that the signified has a natural bond with the referent, and that while different signifiers are used in different languages, the signifieds to which these signifiers are correlated are the same for everyone across all cultures. (Bennett 2021: 11)

Hjelmslev is a perfect example of anti-logocentrism. What Hjelmslev is trying to say with the *slon* example is that there is no natural bond between the signifier and the referent, signifieds are different across cultures. Signifieds are different across cultures in terms of substances, but what about forms?

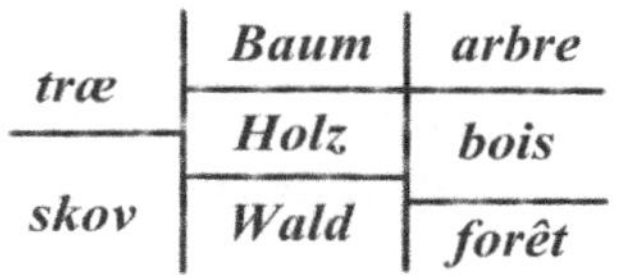

træ	*Baum*	*arbre*
	Holz	*bois*
skov	*Wald*	*forêt*

Figure 24: Form of substance according to Hjelmslev.

Let´s have a look at Fig. 24. This famous schema from *Prolegomena* illustrates the form of the content, again, with the aid of interlingual translation between Danish, German, French. Here we see that in Danish, *trae* covers all of the German *Baum* and the French *arbre*, and partly covers the German *Holz* and less of the French *bois*. Similarly, *skov* partly translates the German *Holz* and *Wald*, as well as most of the French *bois*, and some of the French *foret*. This table illustrates that not only substances but also forms are different across languages. The crucial question is that if the substance of a language must be related with other social institutions or cultural artefacts, as Hjelmslev proposed in *La stratification du langage*, how can it be actually determined or subdivided by purely abstract form? I can see here an inconsistency between what is claimed in *Prolegomena* and what is narrated in *La stratification du language*.

When we relate this understanding of the form with the hierarchy between form and substance, the question becomes: if the form is elusive and subject to change, how can it be in a dominant position towards the substance? The essay *Langue et parole* is where Hjelmslev theoretically describes this elusive character of the form. In this essay Hjelmslev takes the Saussurean dichotomy *langue* (form) – *parole* (substance) and divides it into three terms: scheme, norme and usage. Scheme is what most corresponds to *langue*, the pure form, whereas norm is something like habit taking while becoming scheme and usage is the pure use of language, analogical to *parole*. Already the act of transformation of the dichotomy to a trichotomy is significant in connection to Peirce. But what is more important in the scheme-norme-usage trichotomy is the order of the determination of the terms among themselves. Hjelmslev proposes in this essay that it is actually usage influencing the norm which itself determines the scheme and not the other way round.

Argentinian linguist Luis Jorge Prieto also comments on this not-very-clear position of Hjelmslev toward the substance. In his book *Pertinence et pratique* (1975) Prieto dedicated a chapter to Hjelmslev's easy *Langue et parole*. On one hand, he defines Hjelmslev as antisubstantialist, but on the other hand he claims that it is only seems that Hjelmslev avoids substance (see also Chávez 2022: 17–19). The seeming avoidance of substance becomes transparent with the commutation test. Indeed, commutation is the modification of substances within one form serving to differentiate between variants and invariants. Here, Prieto is right that the avoidance of substances is illusory, maybe not only in commutation but in the whole of Hjelmslev's work.

I would like to go in the direction implied by *Langue et parole* and continue to theorize about the dominant role that substance can take over form. Only by attributing the dominant role to substance, reversing the order of hierarchy between form and substance as proposed in *Prolegomena*, are we able to use the notion of form in science – especially in biology – and at the same time avoid Platonic idealism and accusations of vitalism. This is also a way of giving credit to Peirce's essays on organic growth where he uses the notion of form abundantly. Epigenetics also reverses the order between form and substance in theoretical descriptions of the living. If we understand the genetic script as the *form* with potentiality to create substances (formed living beings with inherent morphology) within the epigenetic view and the Extended Evolutionary Synthesis, there is room for this reversal.

The notion of "norm" is also applied to biological phenomena by Markoš and Švorcová (2019: 180). "Norm" overlaps between variants, guaranteeing mutual understanding among lineages. Markoš and Švorcová prefer the term "norm" over the term "code", because norm is constantly being negotiated; it is symbolic, yet not stuck or frozen. We can speak about norm with the definite article and with the majuscule N, as the authors do in the monographs (2019), saying that living beings develop the Norm endlessly. According to the authors, the Norm is necessary but incomplete and it is being completed dynamically by organisms and their environment:

> The Norm included, not only basic metabolic and genetic processes (those that could have been established in the prebiotic phase), but also a principal novelty that could be introduced by cellular life alone – the exchange and processing of information, both within the cells and from outside. (Markoš and Švorcová 2019: 96)

The Norm is what we should understand by the genetic code as matching between amino acids and nucleic bases *plus* a bunch of additional factors (the genetic code itself alone can be defined as a scheme in Hjelmslev's terminology, a table of matches between amino acids and triplets of nucleic bases). Markoš and Švorcová hypothesize that the current Norm is more complex compared to the original Norm of the first cells because of being constantly negotiated, narrated and transferred to subsequent generations with potential modifications:

> Norm that took grip with cells: it embraces, e.g., five nucleotides, twenty amino acids, a handful of phosphorylated sugars, etc., all with the proper molecular handedness (chirality). The first cells started building biospheres with such a humble set – essentially, with what we find in the basic biochemistry and cell biology textbooks. This is the underlying "logic of life". (Markoš and Švorcová 2019: 3)

Epigenetic marks constitute additions to the Norm; existing among the genetic (generic) one, they are both "norms of symbolic communication with the world". Within Hjelmslev's original trichotomy of scheme-norm-usage, norm is the transitional step of habit making, establishing a law. Norm can answer the problem of the tension and complicated relation between chance and law in Peirce. As habit making in Peirce, norm in Hjelmslev is the most interesting part of the terminology for the description of the evolution of species or the evolution of languages. Norm is the area where interpretation and creativity take place, the negotiation being done between language users or among organisms. It is not lawless, though it lacks strict rules. Interestingly, "Habit" in CD, p. 2673 (25/ 2/2024) is defined as "usage", thus even not-yet-Norm. Under meaning n. 1, beyond other meanings, "development" is emphasized. It is more than explicitly said in many Peirce's papers that habit is a matter not only of our mental action (most often unconscious, even involuntary), but also of all living beings, and not only living. Peirce associates habit also with inanimate matter, but I leave this out of my argument here.

Habit is the fixation of beliefs – it is not by logic but by habit that we form our conclusions.

That which determines us from given premises, to draw one inference rather than another, is some habit of mind, whether it be constitutional or acquired. (Peirce 1877 4)

Returning to the notions of form and substance, let me again quote the CD, p. 2335 (accessed 25/2/2024) definitions of both terms, specifically meaning n. 8 of the entry "Form":

arrangement of relationships between the parts of anything, as distinguished from the parts themselves: opposed to matter, but not properly to substance. Thus, to say that the soul was immaterial was formerly considered the same as to say that it was a form. With the older writers form is often synonymous with essence, and has generally lofty associations (thus, the shape of a living being, considered as its perfection, was called its form), and these ideas cling to the word in the minds of later writers as Kant. But with many modern writers the conception is of something imposed upon the thing from without, and distinct from its life and essence. (CD: 2335, accessed 25/2/2024)

After that, the entry comments on Plato and Aristotle's understanding of form, which I already mentioned. Let me call attention to the understanding of form as "opposed to *matter*, but not properly to *substance*". This detail in particular, to differentiate between substance and matter, is extremely Hjelmslevian. In Hjelmslev, substance is also defined as formed matter; that is, in Hjelmslev's terminology, it is opposed to "purport," and is not an amorphous mass.

"Substance" is defined as "that which exists by itself," (CD: 6030, accessed 25/ 2/2024) and is also related to *essence* as in the entry for form. This means that both form and substance are related to the notion of essence, and it is because of this fact that essence is understood by Peirce and defined in CD as form, and substance is but matter accustomed to form.

It might seem that the form-substance bipartition is only reactionary thinking in terms of the structuralist binarism. Still, form and substance are not important for Peirce's theory of relations (semeiotic), because it is inherently formal by its very nature. Form and substance were nevertheless very important notions for his observations of nature and what he called the "law of nature", as well as for his theory of the phaneron. This is what Peirce and Hjelmslev have in common: considering form means also considering *substance*. In other words, we cannot give the only and unquestionable importance to *form*. Form and substance are what make Hjelmslev special in comparison to other first generation semiological theories. The classical structuralist theories, if we take the Geneva school, by definition treat language formally: structuralism derives from structure, which is nothing but form. The addition of the form-substance dichotomy by Hjelmslev creates a completely different approach. It implies that there is *something beyond* the form, an observation which is lacking in approaches without the notion of substance, that is, without the form-substance dichotomy. Moreover, as I already mentioned, Hjelmslev expressed in *Langue et Parole* somehow more explicitly the possibility of the primacy of substance over form. This discussion about the substance is extremely relevant for biology, regardless of the fact that there are forms in biology and as Peirce described them, these forms are not Platonic forms. But it is important to remember here that forms are only possible in relation to substances. Above all, in the CD biology is defined as a science of substances by Peirce (entry Science: 5397, accessed 14/1/2024).

Regarding the status of substance, the most important notion from Hjelmslev's theory is *participation*. It was originally used in anthropology and applied to linguistics by Hjelmslev. I will now apply the notion of participation to biology, as one more tool to demonstrate the inseparability of nature from culture and the shared pattern among the biological and the cultural. This is to repeat that there are the same relations, the same patterns in nature and in culture in the sense of Peirce's semeiotic as relational science. According to the NonReduction Theorem, the structure of a protein (3D shape) is not reducible to dyadic relations. The peptide sequence as a chain composed of dyadic relations will never sufficiently express the 3D protein structure. One might argue, however, that protein structure *is* reducible to dyadic relations. One cannot deny it, because the existence of the peptide chain is proof. My answer to this argument is that, as already mentioned previously, the peptide chain is rather a case of the *generating of* dyadic relations and not of the *reduction to* dyadic relations. In line with the theory of folding the continuum, I maintain that dyadic relations can be generated from a triad, yet cannot themselves generate triads. It is true that in the process of protein synthesis, a protein is construed by the linear text written in amino acid language, which could be understood as a generation of triads from dyads. One should not, however, get confused. From a semeiotic perspective, the process would be described as follows: the dyadic chain has to first be generated, meaning *articulated*, from a triad (since before it was only a continuous unarticulated line) and only after can it turn out to be a tool for creation of other triads. This is still not a completely correct vision of the thing. It is true that the peptide chain serves as a starting point in protein synthesis, but it is not the unique factor of creation of the final protein's shape. The *context* must finalize the entire process. We can admit that the dyadic chain generates somewhat the triadic protein, but it only generates a degenerate triad[34], which is to be completed by contextual factors.

---

34 "A Relation is either Genuine or Degenerate. A Degenerate Relation is a fact concerning a set of objects which consists merely in a partial aspect of the fact that each of the Relates has its Quality" (CP 2.91).

Generating is not, however, the most correct term. The peptide chain is a medium to express organic meanings: likewise, language is a medium to express linguistic meanings. This does not mean that language generates meanings (although this can be affected in some generative grammars). The very problem of the limitedness of a linear string in expressing non-linear meanings is easily demonstrable by ambiguous sentences, as in the Italian example (La vecchia porta la sbarra)[35], which can be interpreted in two ways depending on whether we consider the word "vecchia" as a substantive (an elderly lady) or as an adjective (old), then accordingly if we consider the word "porta" as a substantive or as a verb "portare", the word "la" as an definite article or a personal pronoun etc. The two interpretations can be schematized consequently:

(1) La vecchia porta la sbarra

    a. $[[La_{det}$ vecchia$_n]np]$ $[porta]v$ $[[la_{det}$ sbarra$_n]$ np]$

    "The old lady brings the bar"

    b. $[La_{det}$ $[vecchia_{adj}porta_n]n]np$ $[[la_{pron}$ sbarra$_v]$ vp]$

    "The old door bars her"

One can argue that my example is barely valid, since the sentence *La vecchia porta la sbarra* is an example of a very rare linguistic phenomenon, an amphiboly, wherein several polysemantic words happen to be placed together by accident, and consequently, my example is a language rarity rather than a general linguistic trait. But, as a matter of fact, polysemy is more present in the everyday use of language than one might think. Most lexical units (and in particular those with high frequency of usage) are polysemantic. For instance, the majority of highly frequented verbs are polysemantic. The English verb "to carry" may acquire many different meanings depending on what words are in its proximity: carry about, carry along, carry away, carry forth, carry forward, carry back, carry in, carry on, carry off, carry over, carry up, etc. And the polysemantic nature of the verb is not exclusively related to phrasal verbs. According to the Oxford English Dictionary (Stevenson 2010), there are even more than forty-three meanings of the verb "to carry" which are not phrasal uses of the verb: to transport, to bear or take (a letter, message, report, news, and the like), to take by force, to cause to go or come, to extend or continue (a line, a piece of work) in the same direction to a specified distance, to win, and many others.

---

35   This phenomenon might be also considered as an example of the so-called bracketing paradox (Spencer 1988).

Peirce was aware of the limits of the linear representation of relations, which is also the case for language because of its characteristic of being linear, which was why he proposed a complex tool for nonlinear expression of logical relations (the Existential Graphs). Linear strings have, in contrast, an enormous advantage, having the power to encode and store for ages a huge amount of data with only a few letters (alphabets of natural language have around 26 letters, the amino acid alphabet has 20 letters). Linear strings are very economical and practical. The price for their economy is the fact that they are never exhaustive: an expression never exhaustively encompasses the meaning it refers to. Of course, exhaustiveness is even undesirable: at the moment when the expression exhaustively expresses the meaning, the two would equal, and thus the very role of expression would lose its sense. Take an example of a map as a representation of a given territory. A map is a special type of representation, trying to match with the territory of the real terrain to the greatest possible extent. A perfect map does not exist, since a perfect map would have to represent every single smallest point of the territory. It would have a dipstick of 1:1, and hence it would lose its justification and cease to be a map. Thus, the limitedness of expression is a desirable characteristic and the very nature of language lies in it. An expression cannot exhaustively encompass the meaning, but the meaning can exhaustively encompass its expression. A protein encompasses every one of the amino acids of the peptide chain and the bonds between amino acids do not break. It encompasses its expression, but has some additional value, having a biological function given by the shape and by the context. The territory encompasses every single point on the map, but also has some additional information. Apart from the points on the maps, it also has many more additional points: people walking on the streets, some insignificant details, street art, trees, etc. This is the fundamental relation between content and expression in every semeiotic system. Content encompasses its expression – it is a matter of a participative relation or participative opposition. This is the way Paolucci explains the relationship between content and expression: with aid the of participative opposition as introduced by Hjelmslev (Paolucci 2010: 351).

What Does It Mean to *Participate in*?[36]

Louis Hjelmslev borrowed the term "participation law" from the anthropology theory of Lévy-Bruhl and elaborated it in a more linguistic way as the constitutive

---

36   This chapter has been partially published as Lacková, Ľudmila (2022). Participative opposition applied. *Sign Systems Studies*, 50(2–3), 261–285.

character of language. He developed this theory primarily in two essays: *La catégorie des cas* (Hjelmslev 1935) and *Structure générale des corrélations linguistiques* (Hjelmslev 1985). The core idea of the participation law governing a linguistic system resides in that, according to Hjelmslev, language is not analyzable in terms of binary oppositions, as opposed to the mainstream of structuralist linguistics of the second half of the 20th century; nor is it analyzable in terms of *exclusive* binary oppositions. Hjelmslev himself, being one of the most influential structural linguists, did not deny binary relations in language, but did conclude, in a kind of compromise, that the binary oppositions governing language are not of an exclusive character (as in the mentioned essay *Langue et Parole*). In other words, they do not exclude one another. To put it more simply, this means that the terms in the participative opposition may coexist without excluding one another. Not all linguistic categories are definable, of course, in terms of participative oppositions. In phonology for instance, oppositions are always exclusive[37]: one phoneme cannot be anterior and posterior at the same time, or labial and non-labial at the same time. The impossibility of superposing two contradictory features (distinctive features) at the same time guarantees the definition of a phonological unit: a phoneme is defined by exclusive oppositions, by phonemes with which it is in opposition. /p/ is /p/ because it is not /b/, with which it creates an exclusive binary opposition. /n/ is /n/ because it is not /ŋ/, with which it creates an exclusive binary opposition. When we move, however, from phonology to "upper" linguistic areas, such as morphology or lexicology, the definition of linguistic units by means of exclusive oppositions becomes more complicated. It would be limiting to define, for instance, the word man by an exclusive opposition with the word woman, since the meaning of the word man is not limited to an opposition with the word woman. Imagine the following sentence: "All men are wise". In this case, the meaning of the word *men* is not definable by opposition to *women*. Hjelmslev observed, however, that in some way, there is a relation of opposition, and resolved it by calling this kind of opposition a participative opposition. The term women is in opposition with the term men, but at the same time is *included* within the term men. This paradoxical situation is, according to Hjelmslev, by no means an exception within the language system and it is not

---

37   It should be remarked that, according to the phonological theory of the Prague school, the exclusiveness of the phonological oppositions is not that obvious. The Prague scholars suggest treating some of the phonological oppositions rather in a scalar manner than in terms of exclusive opposition. As a consequence, they differentiate between privative, gradual and equipollent oppositions in phonology. (Trubetzkoy 1939: 67–74)

merely a particularity of semantics or lexicology. In a similar manner, all morphological categories are definable by participative oppositions. The case system, verbal tenses, gender and number of substantives, all these categories enter into participative opposition. As a result, the accusative is opposed to the nominative, but is included within it at the same time. The plural is opposed to the singular but is also included within it. The past is opposed to the present but is at the same time included within it, etc. How is the accusative included in the nominative, the plural in the singular and the past in the present? Firstly, it is by the syncretism of different functional units (nominative and accusative) in one morphological form (nominative) within a given paradigm. Secondly, it is by syncretism of different meanings (past, present) in one morphological form (present) in a specific context (the historical present for instance). In the case of the plural and the singular, one can also speak of metonymy. Thus, participative opposition is, according to Hjelmslev, a constitutive principle of language.

The above-mentioned examples of a participative opposition between cases, tenses and the grammatical category of number illustrates the way language disobeys the laws of classical logic. By participative opposition, contradictory features coexist, with one and the same unit being accusative, vocative and nominative, all at the same time, without losing their identity. A linguistic system is free when compared to the logical system that corresponds to it. On the axis of the logical system it can be oriented differently, and the oppositions it creates are subject to the law of participation: there is not an opposition between A and non-A, the only oppositions in the linguistic system are between A on the one hand and *A + non-A* on the other hand (Hjelmslev 1935: 214). Hjelmslev points out the illogical nature of the linguistic system, a system in which the basic prerequisites of Aristotelian logic (the identity law, the law of the excluded third) do not work. Such a system is called a *sublogical* system. Hjelmslev claims that only a sublogical system can describe language phenomena. The core of the work of Hjelmslev lies in the opposition between an intensive (precise) term and an extensive (vague) term.[38] The intensive term is decided while the extensive term is undecided.

The observations from Hjelmslev's participative opposition lead us in a direction already intimated by the theory of the folded continuum: a term undecided (triadic) on the level of system (*langue*) may become decided, that means dyadic

---

38 More about the problem of participative opposition, its importance for Roman Jakobson and the Russian linguists, and its application to contemporary biology can be found in Lacková 2021.

on the level of language in actual use (*parole*). Thus, the contradictory nature of a linguistic system, where units exist having more and even opposite meanings at the same time (*man* or present tense or singular number or nominative in cases), becomes unambiguous at the moment of speech, of the *use* of language. The moment of speech is a concrete *fold* of the continuum. Only when the continuum is folded, a particularity, an identity arises. At the moment of the folding, ambiguity disappears, man is masculine (A) or it is both masculine and feminine (A + non-A), but it is one of the two possibilities, not two at the same time, not at the moment of speech.

In summary, the seemingly desperate situation of the ambiguity of the linguistic system outlined by the participation law is elegantly resolved when one moves from the system (langue) to the linguistic use (parole). We are slightly touching at this point the field of pragmatics.

Consubstantiality is another term introduced by Lévy-Bruhl, who assigned the feature of participation to language because he was convinced that language makes up part of culture and ultimately that culture is connected to the mind. In this manner, he spoke of the "primitive mind" in primitive societies. He explained the difficult attitude of primitive people towards their own images with the notions of consubstantiality and participation:

> When a savage sees his own image (shadow, reflection, etc.) it is not a more or less faithful reproduction of his features, it is the consubstantiality that he imagines and feels between them and him. But he can also imagine and feel this essential participation between him and a being whose external appearance is different form his own. (Lévy-Bruhl 1996)[39]

Consubstantiality is not only a concern of the visual perception of images in primitive societies – it is a phenomenon that goes far beyond perception. It goes to the very mental processes and is also reflected in the language of primitive societies. As an example, the "primitive" denomination of things in the world even becomes incomprehensible to our understanding. In the culture of Australian aborigine people, for example, the name for the sun and the name for a white cockatoo are considered to have one shared meaning. This is not a simple synonym: however, the fact is that the very signification of the word is

---

39   In the original: "Le primitif quand il voit sa propre image (ombre, reflet, etc.) ce n'est pas la reproduction plus ou moins fidèle de ses traits, c'est la consubstantialité qu'il imagine et qu'il sent entre elle et lui. Or, cette participation essentielles, il peut aussi l'imaginer et la sentire entre lui et un être dont l'apparence extérieure est autre que la sienne.".

the sun and the white cockatoo *at the same time* and without being a case of pure randomness, as is the case of synonymy, nor is it a case of mere polysemy, where the same expression is used to refer to different meanings diachronically and conceptually related ("crown" for instance). The concepts of the sun and the white cockatoo are, in this culture, consubstantial. In other words, for the aborigine people, a situation where something is the sun and at the same time the white cockatoo, is absolutely normal.

For aborigine people, the terms sun and white cockatoo do not exclude each other (Harris 2009). Lévy-Bruhl was primarily dedicated to anthropological studies and not linguistics, but Hjelmslev succeeded in applying the notion of participation to the linguistic system in general. Hjelmslev considered not only primitive languages, but also modern European languages to be governed by the law of participation, whose main trait is that it does not correspond to the laws of classical logic. As a matter of fact, language makes up part of culture and there is continuity between the language of indigenous people and modern languages. As a consequence, all modern languages carry the heritage of primitive languages and as a consequence, all grammatical systems are far from being logical. Moreover, Lévy-Bruhl admits that the difference between the primitive mentality and ours is far from absolute. Every human mentality is marked by a certain primitiveness. The representations and the connections of the representations governed by the law of participation are far from having disappeared (Hjelmslev 1928: 262)[40].

The law of participation, characterized by consubstantiality, disobeys the classical logic law of the excluded middle. As a consequence, the sun is at the same time the white cockatoo, the future is at the same time the present and the nominative is at the same time accusative. There is nothing inconsistent; however, in this observation, modern languages have simply inherited certain features from primitive cultures. This is how Hjelmslev explains the illogical nature of language.

In proteins, consubstantiality is more evident than in natural language, with the linear peptide chain clearly sharing its substance with a functional protein; but if the two parts of the semeiotic function (expression and content, or, peptide chain and protein) both partake of the law of participation, what is the formal

---

40   In the original: "D'ailleur, M. Lévy-Bruhl admet que la différence entre la mentalité primitive et la nôtre est loin d'être absolue. Toute mentalité humaine est empreinte d'une certaine primitivité. Les représentations et les liaisons des représentations regies par la loi de participation sont loin d'avoir disparu."

difference between the two? The only possible answer is double articulation, the concept introduced by French linguist André Martinet (Martinet 1967), and which is extended by Deleuze and Guattari. Double articulation actually treats one and the same substance as disposed to be articulated (folded) in two steps, where the first step affects content (articulation of phonemes, which, by combinatory rules, create words and sentences) and the second step affects expression (articulation of single phonemes). In the organic context, the first articulation affects the folding of the peptide chain and the second articulation affects the individuation of single amino acids. The distinction between the two articulations is not, as pointed out by Deleuze, between forms and substances, but between content and expression:

> A stratum always has a dimension of the expressible or of expression serving as the basis for a relative invariance; for example, nucleic sequences are inseparable from a relatively invariant expression by means of which they determine the compounds, organs and functions of the organism. To express is always to sing the glory of God. [...] The first articulation concerns content, the second expression. The distinction between the two articulations is not between forms and substances but between content and expression, expression having just as much substance as content and content just as much form as expression. (Deleuze and Guattari 1987: 43)

Deleuze's understanding of the terms content and expression is traditional in the Hjelmslevian sense, in that he comprehends the two sets of semeiotic relation as being interdependent: in other words, only being definable by reciprocal solidarity and in no other way (having no identity as single entities, but only in relation to each other). Deleuze's understanding of the terms content and expression is untraditional, however, in that the two sets for him do not have separate forms and separate substances. For Hjelmslev, in fact, four separate units were defined ((Hjelmslev 1963):

- form of expression
- form of content
- substance of expression
- substance of content.

According to Deleuze, however, expression has just as much substance as content and content has just as much form as expression. This means practically that there is no such difference between content and expression as regards the form and the substance; or, at least, there is no quantitative difference between the two sets, inasmuch as expression has just as much substance as content and content has just as much form as expression. One might then ask, however, what the difference is between content and expression? According to Deleuze, this difference is only and exclusively in the articulation. The difference is in the way of articulating the formed substance. By articulation, in the case of protein folding, I mean the singular folds, the folding of the continuum.

The difference guaranteed by articulation is important since the first articulation is much more complex than the second articulation. There is never a correspondence or conformity between content and expression, only isomorphism with reciprocal presupposition. The distinction between content and expression is always real, in various ways, but it cannot be said that the terms preexist in their double articulation. It is the double articulation that distributes them according to the line it draws in each stratum; it is what constitutes their real distinction (Deleuze and Guattari 1987: 44).

The "line" Deleuze refers to is the "cutting line" that cuts the continuum – in our case it would be better to talk about "a folding line", as I proposed previously, inasmuch as the transition between organic strata is given by the fold of the peptide chain. Deleuze's observation that "it cannot be said that the terms preexist their double articulation" and that "it is the double articulation that distributes them according to the line it draws in each stratum" leads us back to the impossible preexistence of the expression without a connection to the content. It is thanks to double articulation that the consubstantial linear chain appears both at the level of content and at the level of expression. The analogy between the phonic chain and the peptide chain is very clear in this case: the first articulation – protein folding – affects the content, and the second articulation – the individuation of single amino acids – affects the expression.

At this point it is worth mentioning the work of Ji, who elaborates in detail the application of the concept of double articulation to protein folding. The first articulation is also attributed to protein folding by Ji, and the second articulation is attributed to single amino acids forming the peptide chain:

> Also called double articulation, duality is considered one of the most fundamental characteristics of all languages [...] The first articulation is responsible for the almost infinite number of sentences that can be generated from a finite number of words, obeying a finite set of combinatorial rules called grammar. The second articulation refers to the formation of words by combining simpler units, phonemes [...] and morphemes [...]. Cell language also possesses duality, and I attribute this ultimately to the duality of molecular interactions, namely, covalent (also called configurational) and non-covalent (or conformational) interactions. Examples of covalent interactions include linking nucleotides or amino acids to form nucleic acids or polypeptides, respectively, and phosphorylation and dephosphorylation reactions in signal transduction pathways. Examples of conformational interactions are provided by folding reactions of biopolymers and binding reactions between ligands and receptors and between transcription factors and DNA regulatory sites. (Ji 1999: 24)

It is apparent that Ji's usage of the notion of double articulation is absolutely coherent with the linguistic tradition, the first articulation being attributed to larger languages units like sentences and the second articulation attributed to smaller language units like phonemes (or morphemes). Analogically, applied to proteins, the first articulation is attributed to proteins, and the second articulation is attributed to amino acids. With the aid of Deleuze's treatment of double articulation as concerning the content (first articulation) and the expression (second articulation), we can conclude that in the case of protein folding, it is semeiotically convenient to approach functional proteins as contents and their peptide chains of amino acids as expressions. I specify protein folding because in another case the situation might be different: semeiotic units are not frozen entities, but always definable and redefinable in accordance with relations and context. Moreover, the discussed notion of double articulation can be connected to the theory of folded continuum: the first articulation is represented by the folding of the linear continuum (peptide chain) while the second articulation is represented by the binding together of basic units constituting the very linear chain (amino acids). Note that also in the case of Martinet's double articulation, and similarly to the Peircean theory of continuum, the direction goes from non-linearity to linearity; in other words, from triadicity (dimensionality) to dyadicity (flatness or linearity). The more complex units are defined first, and only consequently do the basic units emerge.

The tension between linearity and three dimensional complex relations is present both in language and in protein folding. In language this is the tension between the linear speech chain and the irreducible multidimensional linguistic meanings. In molecular biology, this is the case between the peptide chain and the three dimensional protein with its many biological functions. The analogy between language and the peptide chain in is my contribution to Jakobson's original language metaphor of DNA. The law of participation explains the problematic question of how the passage from a linear string (sound) to a complex structure is impossible to express linearly (meaning). The meaning of a sentence, for example, is a complex structure impossible to define by means of linearity. Even if the sentence is expressed by a linear phonetic chain, to understand the complexity of the meaning of the sentence (all relations between the single constituents) a non-linear syntactic representation is needed. One can therefore argue that the phonetic chain is encompassed within the meaning of the sentence: it *participates in* the meaning. The phonetic chain obviously participates in the meaning, but is not sufficient to decipher the entire meaning, other factors having to be added such as syntactic relations and pragmatic factors. To put it in a different way, expression is a condensed version of the meaning, it is the intensive term. This approach allows us to comprehend the semeiotic notions of content and expression not as two separate entities, but as being part of each other, indissociable, as the celebrated metaphor of de Saussure states, two sides of one sheet of paper. Thought is the front and the sound the back; one cannot cut the front without cutting the back at the same time (de Saussure 2011: 113).

Yet the law of participation goes beyond the metaphor by de Saussure. The metaphor is, as a matter of fact, misleading, since it implies that content and expression are equal entities in the sense that both sides of a paper sheet are of the same size and material. Paolucci's idea of applying participative opposition to the relationship between the expression and the content, however, shows the asymmetry between the two. Content and expression are not of the same rank or the same degree. The participative opposition between content and expression consequently shows reciprocal unsociability on the one hand, and asymmetrical inequality on the other.

The participative opposition, being also a kind of binary opposition, creates a compromise between strict binary structuralist thinking and semeiotic theories that deny the binary character of semeiotic phenomena. I believe that what nature and culture share is the diagrammatic pattern consisting in participative opposition between linearity and nonlinearity. In other words, nature and culture are both definable in terms of sublogical systems. The sublogical nature of human societies is described by Peirce in his *Fixation of Belief*:

> We are, doubtless, in the main logical animals, but we are not perfectly so. Most of us, for example, are naturally more sanguine and hopeful than logic would justify (Peirce 1877: 3).

Peirce's inclination to modify classical propositional logic is clear already from his EGs, most strikingly from the Gamma system. He was convinced, similarly to Hjelmslev, that in language (but not only in language), there is a pattern, but its logic cannot be expressed linearly or symbolically. A diagrammatic iconic notation is needed to describe the sublogical nature of semeiotics. The participation law is present in Peirce under his NonReduction Theorem. The impossibility of reduction of the triad is but another definition of the participation law: dyads participate in triads, but the latter cannot be reduced to the former.

On Writings and Alphabets

The language metaphor of life was born from the deciphering of genetic code, which led scientists (and linguists) to believe that our biological nature is determined by a binary code, the *form* of an unsubstantiated virtual script. Thanks to the theory of participative opposition applied to the form-substance distinction, and thanks to the retroaction of the substance upon the form (Bennett 2021), we can see the limits of such understanding of life determined by the virtual script. The language metaphor of life has been recently critically approached precisely because of this belief (Rączaszek-Leonardi and Deacon 2019). The criticism is based on Terrence Deacon's model of gradual language emergence in the human species from iconic to symbolic communication, which contradicts the usual understanding of the genetic code as a digital symbolic system. In these approaches, language is understood as embodied or distributed, thus de-centralized from the human brain. Here I propose that the tradition of the language metaphor of life is based upon a wrong understanding of what language is, but the main issue is not the emergence of symbolic communication. The main issue in the language metaphor of life is how language relates to alphabetical writing. The problem seems trivial, but I will explain how it is crucial to focus on the question of alphabetical writing in the project of *decentering* (in the Derridean sense) biology away from genetic determinism.

Since the deciphering of genetic code, many questions have been posed about the accuracy of the language metaphor, including whether it is a pure metaphor or if the genetic code is a *real* language. Yet one peculiarity of the language metaphor of life has not been openly addressed: the metaphor is not so much with *language* as it is with alphabetical *writing*. Genetic bases are described as letters and the RNA strings are *transcribed*. Because of this, I believe, many (especially) linguists cannot see more than a mere metaphor in the analogy between the genetic script and human natural language. This is understandable because modern western linguistics focuses on the study of spoken language, – written language and writing systems are studied by other disciplines (history, anthropology, cultural studies). Writing is considered, from the perspective of the linguist, something inferior – it is a mere transcription of what the "real" pure language is. This understanding of language as primarily spoken is supported by the development of linguistics in the last decades towards communication analysis, pragmatics, cognitive linguistics and discourse analysis with the use of language corpora.[41] The categories used in linguistics, especially in phonology, are nevertheless dependent on written language. Phonology postulates that phonemes are the minimal elements of linguistic analysis. This postulation has never been questioned because we have alphabetical writing so engraved in our minds that it is unnatural for us to question the rule that a single letter corresponds to

---

41 A great review of this discussion is magnificently summarized by Derrida in the first chapters of his *Of Grammatology* (1967).

single speech sound. This is also why it is difficult to apply categories of western linguistics to languages with pictographical writings. Linguists despise writing, but unconsciously are influenced by its rules. The language metaphor of life can be real, but we need to liberate ourselves from limiting language to alphabetical writing. In the same way as language is *not* the alphabet (it is much more, it is the poetry, the tone the mother speaks to the baby, the multidimensional ideas and emotions expressed verbally and non-verbally), the genetic script is not the code of four letters. The language metaphor of life is about how and for what the language is used. It is about development, evolution, epigenetics.

Derrida uses the notion of *différance*, trace, and arche-writing when he describes the possible primacy of writing over spoken language, but what Derrida means by writing here is something notoriously ambiguous and two-sided. In the first place, what he means by this primordial writing is the possibility for all types of *inscription*, irrespective of medium or substrate. The notion of trace is borrowed from the gravestone rubbing, when one puts a sheet over the cement surface and outlines the text with chalk or crayon. All kinds of enacted speech for example are posterior to the matrix of potential inscription, which is a kind of representational predisposition not only of man, but of all of nature. From this view, writing as *we* know it – the writing that I am doing now, with my keyboard – is posterior to the primordial "writing". In some sense we may connect Derrida's "arche-writing" with the "symbolic" capacity that Terrence Deacon refers to, which is a representational capacity that emerged in evolution long before actual speech appeared (1997), however neither Derrida (nor Peirce) would restrict access to this profound matrix to humans alone, nor even necessarily to animate matter. The hypothesis already smacks of the kind of transcendental or idealistic metaphysics we hope to avoid, but this is why Derrida is so important in this picture – because (in the second place) this "arche-writing" is also an *actual style of very difficult writing*, whose difficulty is supposed to *defer* the linear closure of the symbolic interpretation/inscription, preserving potentiality. This is the intellectual idiosyncrasy for which he and many of the other second-generation semiologists (Roland Barthes, Louis Hjelmslev, and especially Giles Deleuze and Félix Guattari) were so well known; it is also a textual and aesthetic consideration.

But it is nevertheless the very same potentiality that, for example, the POTHOOKS are designed to notate. In the language of the Existential Graphs, this matrix of potentiality also has to do with the "sheet of assertion" itself, before anything is inscribed there (see Stjernfelt 2022). The matrix is *real* prior to its inscription, but does not yet *exist*, to use Peirce's specific meaning for these terms. Needless to say, there are enormous differences between Peirce and Derrida's conception of potentiality – Peirce discovered the matrix by probing the limits of the logical notation and chemistry, and proposed to chart it by means of an enhanced, three-dimensional spatial notation, something akin to today's topology; Derrida's method for exploring the same matrix was to expose the limits of linguistic conceptualization, by means of intertextual transposition and interlingual translation: in between the texts, the trace of something more profound appears. Derrida's celebrated challenge to the *alphabetism* of writing is meant to postulate an idea of language freed from *linearity* and restored to *potentiality*, and it is only this "language" that is suitable for the language metaphor or life.

This book has not consisted in any thorough discussion of the methods of deconstruction or second-generation semiology – it has rather been an excursion into the status of genetics and biology from a Peircean point of view; but this excursion reveals the major questions of biology and genetics to be unanswerable in the reductive language of contemporary scientism. Specifically, only when we treat the *potential* as *real* can we account for the natural transformations that interest biologists so much.

As modest as this may sound, in fact it entails a scientific revolution so extensive that even some of the most obscure practices – such as deconstruction – turn out to have explanatory power in the natural world. Without doubt, the much maligned qualitative methods of the humanities – their subjective self-reflection and space making for internal logical incompatibility – are clearly required for any persuasive account of biology. The major claim of all of Anton Markoš' work, cited so frequently in this book, is just this need for hermeneutics and interpretation to describe biological realities. As argued above, Peirce made the same claim, that biology is closer to the humanities than it is to physics or chemistry; however, Peirce himself dedicated little time to consideration of the humanities proper. It is rather the case that what Peirce does, better than any other philosopher, is *prove* with scientific tools and logical argumentation, the importance of the humanities (see also Bisanz 2009: 9–12), even while he does not himself execute the humanistic work that evidently needs to be done. So, on a more local level, we strike this comparison between Peirce and Derrida in order to assert that semeiotic today may rely on a wide variety of theoretical coordinates; and further, to uphold the *homological* method.

The comparison between Peirce and Derrida, regardless of the enormous distance between the two schools they represent, is especially striking if we focus on a certain criticism of linear notation (writing) present in both Peirce and Derrida. Especially reading Peirce's MS 467 dedicated to the Gamma part of his EGs, the reader notices what role diagrams and iconic reasoning played in the whole of Peirce's work. In the introduction to the MS 467 Peirce explains the differences between thinking in language (by "language" here Peirce means linear speech) and thinking in diagrams. He states that a person who learned thinking in diagrams acquires a level of a state of mind of such brightness that for such a mind "it is practically impossible to communicate to the mind of a person who has not that advantage" (MS 467: 4). Peirce is saying here that first of all, our thinking is dependent on the *kind* of language we use and, second of all, which is very optimistic, that it is possible to *train* a person in using different language of thought. For Peirce, language is exclusively linear spoken speech while diagrams are triadic schemes. For Derrida, such a diagrammatic tool for thoughts (but also for communication) was *writing*, and here writing of course is not the alphabetical writing or phonological alphabets. Peirce's diagrams are also, first of all, *written*, that is, graphically represented, this is because our voice (speech) is limited by the dyadic linearity while our hand, when *writing*, is not. I prefer using the term language in a broad sense, where language is not limited to speech nor to phonological alphabets. In doing so, I am opting for a disavowal from the dogma of modern general linguistics considering speech as the only legitimate object of study of linguistics (De Sassure 2011). Linguistics must become such a scientific field which is characterized by inclusivity, therefore for example people without the possibility of producing or hearing speech are language users in the same way as hearing people are. Of course deaf people *are* using language, yet they are not using speech. Deaf people are using language to think (to organize their thoughts), yet they do not have an *inner* voice. I use the term *language* exactly in the same way as Derrida is using the term *writing*. Therefore, "language" in the title of this book, *Language of Life*, is to be language freed from the dictate of the speech. Therefore, the language metaphor of life must not be understood in terms of letters and alphabets. Rather, it should be understood in terms of the relation between linearity and non-linearity, in terms of *folding*. Already Markoš (Markoš and Faltýnek 2011) proposed a re-definition of the classical language metaphor of life, a revolutionary proposal where he actually proposed that the reading of the genetic script is not letter by letter, but rather a more pictographic way of reading. This would imply, to a certain extent, a heresy from the current paradigm. This is also probably why Markoš's idea has not been developed further or more elaborated. It would contradict the importance of the binary genetic code. Markoš is saying that the reading of the genetic script is happening by shapes – that is, the singular linear concatenation of letters is not crucial for the reading. What is crucial, on the contrary, is the overall shape of the totality of the "letters", their spatial non-linear organization. The epigenetic marks on the script, are by consequence, not digits, but they are modifying the shape of the genetic "pictogram". The change in the shape of the pictogram then leads to the change in the final interpretation of the script (Fig. 25).

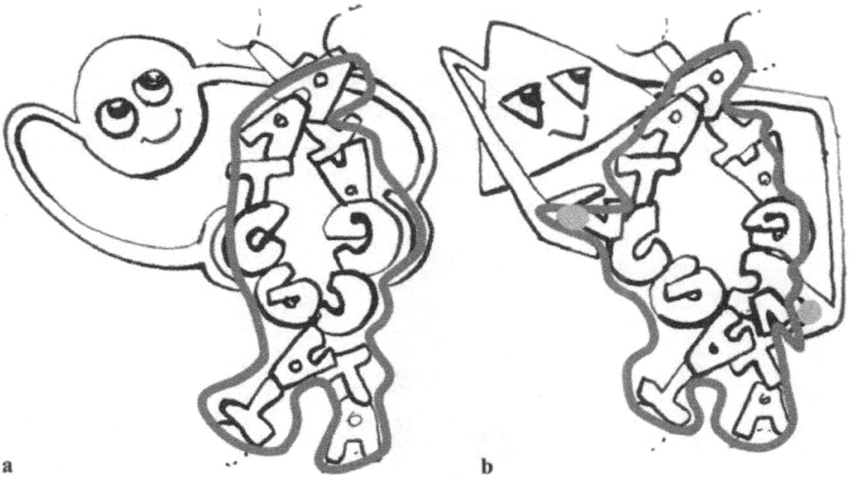

a   b

Figure 25: A genetic pictogram (a) and its epigenetic marks (b). The shape by the pictogram is highlighted in red. The epigenetic marks are highlighted in blue. The shape is slightly altered due to the epigenetic marks. (picture originally from Markoš, Švorcová, and Lhotský 2013, p. 14).

It is our task as the students of the Markoš school of Czech biosemiotics (bio-hermeneutics) to continue spreading this re-definition and investigating its limits and its potentialities. Non-linear graphical representations are probably the writing systems of the future, as was predicted by Derrida. But it is not as simple as saying that oral speech transformed into *a written speech* in the form of texting and writing emails. It is more complex than that. The current system of communication is getting closer to Peirce's Gamma Graphs. Not only is it written rather than spoken, but it is more about multimodality and sharing. As was without doubt clear to Peirce, the existence of something like a collective mind (phaneron) is now getting more fully fleshed out thanks to communication technology.

> By the *phaneron* I mean the collective total of all that is in any way or in any sense present to the mind (…) If you ask present *when*, and to *whose* mind, I reply that I leave these questions unanswered, never having entertained a doubt that those features of the phaneron that I have found in my mind are present at all times and to all minds. (MS 1334, 35)

Today, to communicate means to share. We share news, photographs, memes, emoji, videos, but also documents, on social networks, in private communication, in official communication, in the digital virtual space. The so-called "reality of the virtual" is no longer seriously under discussion when considering the implications of Peirce's non-existing reals, as discussed in this book. VR, XR, etc. are the new forms of reality we live in. Thus, non-linear graphical representations are the writing systems of the future and this future has already begun. But non-linear graphical representations are also the writing systems of the past (Fig. 26). In the jargon of Peirce, semeiosis has no end and no beginning.

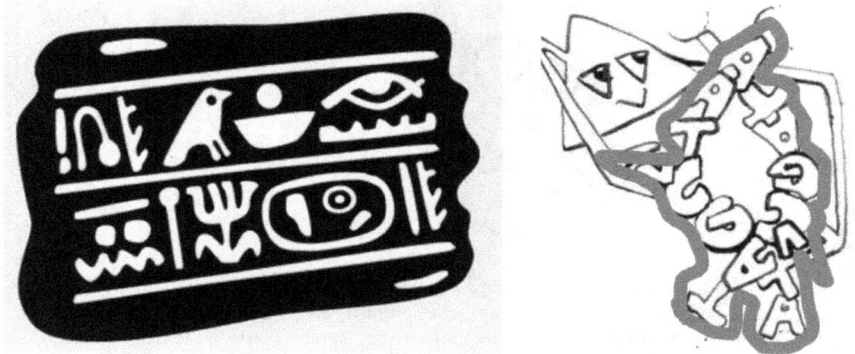

Figure 26: A comparison between a hieroglyph and the epigenetic pictogram.

The language metaphor of life compares (draws a homology between) two non-equivalent domains, and transposes terminology between these two mostly incompatible descriptive systems. It does so not in order to conflate ideas or erase differences, but firstly to expose the inadequacy of either system on its own, and secondly to indicate or "trace" the reality of a more profound system, of which both biology and linguistics are a part, but which itself cannot be exhaustively conceptualized. We can think of no better summary for the semeiotic method. One important intended implication of all this is that the genetic code either *is* this primordial writing, or (probably better) the genetic code is *the first inscription* of this primordial writing. And to recapitulate: the chosen vocabulary of movement between potentiality and actuality is that of *folding*. The folding of a line without intersections (as in the case of DNA molecule folding) is different from protein folding, where the intersections of folded lines create meanings. The folding of a line in a DNA molecule means twisting into a double helix and further spatial arrangements. The DNA molecule in a cell is about two meters long. In order to fit in a minimal space in the cell nucleus it twists and coils and coils again and again. In the same manner, a line of letters folds to fit into a page, and pages fold to fit into a book (Tesnière 1969: 18). A book is a three dimensional object, but the three dimensionality of a book is just an illusion of the folded line, like the DNA molecule is an illusion of the folded line (see Fig. 27)[42].

Protein folding on the other hand is not a simple folding of the line: the folds in a protein are meaningful units. A protein is not determined by the peptide sequence; a protein is determined by its spatial organization. This spatial organization is meaningful: it is not the economization of space. It is the creation of meaning. A folded line creating an intersection is *the* meaning (see Fig. 28).

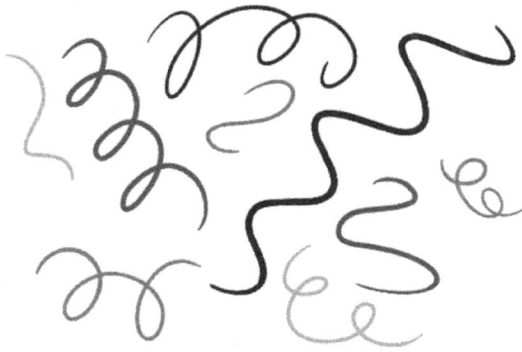

Figure 27: Economization folding without intersection.

Figure 28: Meaningful folding with intersection.

---

42 This is a reductive metaphor. In reality, there are factors in DNA expression which do not follow the linear script but rather its shapes, like in the case of transcription factors.

The possibility of choice between nothing and something happens at the dawn of the binary code. Before zeros and ones there must be a possibility of difference, and this possibility is *real*. Of course, we cannot trace the beginning – semeiosis has no begging and no end.

Afterword: On Minds, Geese and Interpretants

The Institute for Studies in Pragmaticism in Lubbock has been working over the years on a project called Biology of Mind. This book is my contribution to this interdisciplinary cooperation between neuroscience, computer science, AI and Peircean relational logic. The ambition of the project is to describe and account for phenomena from various scientific fields with the aid of Peirce's NonReduction Theorem and semeiotic. The basics of Peirce's semeiotic resides in relational logic applied to the semeiosis model, where the three components, R, O, I represent a triad, that is, the minimal logical relation. In the Biology of Mind the major wedge is introduced as *the interpretant problem*. Often the interpretant is understood as a mental component, an interpreting mind. This is a general mistake. Here I quote a recent paper by Ketner (2022) on the topic of the interpretant within the Biology of Mind project, in semeiotic descriptions of non-self-conscious phenomena, whether physical or biological. According to Ketner,

> he [Peirce] hypothesized that interpreting functions other than those provided by a self-conscious agent could also serve as the interpreting component in a semeiosis. It is sometimes assumed that any candidates for interpreting functions other than humans would also be self-conscious: perhaps an inter-galactic visitor or a fully self-conscious artificial intelligence. (Ketner 2024: 2)

Semeioses that include non-self-conscious Interpretants or [Insc] are in contrast with semeioses based on self-conscious Interpretants [Isc], i.e. human
interpretants. The inclusion of [Insc] resolves the paradox, or the "mystery"
about whether mind can emerge from mechanistic or dyadic relations of the
type action-reaction. Ketner bases his hypothesis on the NRT by Peirce and uses
neuron firing as the prototypical example of such [Insc]. The solution to the paradox of the emergence of mind from dyadic phenomena on lower biological
levels (cellular or neural) resides in applying the NRT to semeiosis and comprehending semeiosis as a constitutive principle of every natural, biological and
cultural phenomenon. This might lead to a misconception of attribution of self-
consciousness to lower organisms or cells or particles. In fact, many biologists
systematically refuse Peirce's terminology because they think it attributes self-
consciousness to cells. In other words, the interpretant is often exchanged for
Interpreter, where Interpreter means self-conscious mind, or something qualitatively comparable to the human mind. As a consequence, for some biologists,
Peirce's terminology is coextensive with a certain kind of vitalism, attributing
consciousness to inanimate matter or lower biological spheres.

In Peirce's theory, the interpretant is but a formal node in the basic triadic
relation of the semeiosis components (R, O, I). Not only does the triadic model
not imply consciousness, but it can even help in the definition of what consciousness is and how it "emerges" from the causal mechanistic constituents of matter.
The difficulties of explaining the phenomenon of mind or consciousness have
been present in human inquiry ever since the birth of modern science. Ketner
says that these difficulties "might be resolved if triadic relational phenomena
are found at the neuronal level, thus allowing a transition to cognition that is
consistent with NRT". In other words, triadic relational phenomena should
be sought at every level of inquiry, even below the "consciousness" threshold.
Ketner's point is that there is no consciousness without a complex neural system (brain), but that triadic relational phenomena do not automatically equal
self-consciousness. On the contrary, Ketner's idea is to contest mechanistic
explanations in science. In recent years alternatives have been proposed, such as
topological explanations. I argue (Lacková and Zámečník 2019) that there is no
basic difference between the widely accepted topological approach and Peirce's
NRT[43]. That is, both are necessarily irreducible to dyadic relations. If we allow for

---

43   See also (Johanson, Arnold, "Modern Topology and Peirce's Theory of the Continuum",
      Transactions of the Charles S. Peirce Society, 37/1 (2001), 1–12. http://www.jstor.org/
      stable/40320822)

the triad as a fundamental relation in all matter, then the paradox of consciousness becomes less paradoxical: we are dealing with triads all the way down, thus no "emergence" is necessary; rather, consciousness is but a semeiosis triad with a particular kind of interpretant, whereas other kinds of interpretant are already present at the molecular level. Yet there is a constant semeiosis from the bottom to the top.

It is hard to account for the triad as a fundamental relation. This is due to the cultural and scientific environment into which we are born. We can blame Cartesian dualism, but alphabetism and dyadic thinking are much older that Descartes. In biology on the other hand, the "notation system" only came very recently, with the discovery of the genetic code. Only in the last century did biology obtain its dyadic "language". At this moment, biology started to be conceived as a real science. Until that moment, biology was classified as "natural philosophy" by Aristotle for instance. Only with the genetic code did biology become "scientific". Consequently, dyadic logic was applied to biology. That is why neurons are understood as mechanistic and why genes are understood as computer programs.

Ketner mentions the example of indexical semeioses displayed by Canadian geese who, when resting by the shore, keep their heads directed towards the wind:

> Some causal features of their sensitive organs make that possible in a manner similar to a wind vane action. Yet the geese are enacting a genuine Index, the Insc interpreting function of which is a hard habit of their system. The geese may not be self-conscious of this habit—perhaps it evolved as a survival adjustment (to escape a predator it is aerodynamically easier upwind to get airborne, as opposed to downwind). (Ketner 2024: 13)

Of course, to our best knowledge, geese are not conscious about their bodies creating indexical signs, in the same sense that the wind vane does not indicate its direction intentionally: it is the human mind interpreting the semeiosis as such. Yet the fact that the geese by this evolutionary habit also escape from predators more quickly implies that the indexicality of this behavior of the species is, even if not defined by the geese as such, functional for them. In the course of evolution, a behavior has been adapted with a teleonomic definition (*in order to* escape predators), which is necessarily a triadic type of relation with R, O and I, the interpretant being the evolutionary habit within a semeiosis that cannot be reduced to the dyadic mechanistic relations of the type "wind blows" → "goose turns her head" → "flying is more effective". To avoid a teleological position we may turn this equation the other way around, so that "for flying to be more effective" → "goose selects the wind direction" → "and positions her body in this direction". I propose, together with Ketner, to apply a spatial triad as a diagram without the need for linear concatenation in either direction. In fact, both mechanistic and teleological explanations share the dyadic demand for linearity. Triadic logic, on the other hand, presupposes no hierarchy of direction, no concatenation – only a triadic structure, where all three components are equally distant from each other.

References

Alberts, Bruce, and Alexander Johnson, et al., "The RNA World and the Origin of Life", in Bruce Alberts, *Molecular Biology of the Cell*. 4th edition (Garland Science, 2002), available from: https://www.ncbi.nlm.nih.gov/books/NBK26876/.

Acosta López de Mesa, Julia, "Peirce's Philosophical Project from Chance to Evolutionary Love", *Discusiones Filosóficas* 25 (2015), 31–41.

Allis, C. D. (ed) (2015). Epigenetics, Second Edition. Cold Spring Harbor Laboratory Press.

Altholz, Josef L., "The Warfare Of Conscience with Theology", in Josef L. Altholz, ed., *The Mind and Art of Victorian England* (University of Minnesota Press, 1976), 58–77.

Ambrosio, Chiara, and Chris Campbell, "The Chemistry of Relations: Peirce, Perspicuous Representations, and Experiments with Diagrams", in Kathleen A. Hull, and Richard K. Atkins, ed., *Peirce on Perception and Reasoning. From Icons to Logic* (Routledge, 2017), 86–106.

Andras, Peter, and Csaba Andras, "The Origins of Life – The 'Protein Interaction World' Hypothesis: Protein Interactions Were the First Form of Self-reproducing Life and Nucleic Acids Evolved Later as Memory Molecules", *Medical Hypotheses* 64 (2005), 678–688.

Ariza-Mateos, Ascensión, Carlo Briones, Celia Perales, Esteban Domingo, and Jordi Gómez, "The archaeology of coding RNA", *Annals of the New York Academy of Sciences* 1447.1 (2019), 119–134.

Beil, Ralph Gregory, and Kenneth Ketner, "*A Triadic Theory of Elementary Particle Interactions and Quantum Computation*" (Institute for Studies in Pragmaticism, 2006).

Beil, Ralph. G., and Kenneth L. Ketner, "Peirce, Clifford, and Quantum Theory", in T. G. McLaughlin, ed., *The Raplh Gregory Beil Memorial Volume: Papers in Theoretical Physics* (Institute for Studies in Pragmaticism, 2012), 1957–1972.

Bennet, Tyler J., *Detotalization and Retroactivity: Black Pyramid Semiotics* (University of Tartu Press, 2021).

Berezovsky, Igor N., Enrico Guarnera, and Zejun Zheng, "Basic Units of Protein Structure, Folding, and Function", *Progress in Biophysics and Molecular Biology* 128 (2017), 85–99.

Berwick, Robert C., and Noam Chomsky, *Why Only Us: Language and Evolution* (MIT Press, 2016).

Bisanz, Elize et al., "Toward Understanding the 'Biology of Mind' Hypothesis", in Elize Bisanz, ed., *Applied Interdisciplinary Peirce Studies* (Peter Lang, 2019), 17–58.

Bolshoy, Alexander, et. al., *Genome Clustering: From Linguistic Models to Classification of Genetic Texts, Studies in Computational Intelligence* (Springer, 2010).

Bolshoy, Alexander, "Towards an Encyclopaedia of Sequence Biology", *Linguistic Frontiers* 1/1 (2018), 65–73.

Brier, Søren, "Biosemiotics", in K. Brown, ed., *Encyclopedia of Language & Linguistics* (Elsevier, 2006), 31–40.

Brier, Søren, "Bateson and Peirce on the Pattern that Connects and the Sacred", in J. Hoffmeyer, ed., *A Legacy for Living Systems: Gregory Bateson as a precursor for Biosemiotic Thinking. Biosemiotics 2.* (Springer Verlag, 2008a), 229–255.

Brier, Søren. *Cybersemiotics: Why Information Is Not Enough!* (University of Toronto Press, 2008b).

Brier, Søren, "Peircean Logic as Semiotic Expanded into Biosemiotics", in Elize Bisanz, ed., *Applied Interdisciplinary Peirce Studies* (Peter Lang, 2019), 59–84.

Brierley, Ian, and Dos Ramos, Francisco J., "Programmed Ribosomal Frameshifting in HIV 1 and the SARS-CoV", *Virus Research* 119 (2006), 29–42.

Brunning, Jacqueline, "Genuine Triads and Teridentity", in Nathan Houser, Don D. Roberts, and James Van Evra, ed., *Studies in the Logic of Charles Sanders Peirce* (Indiana University Press, 1997), 252–263.

Burch, Robert W., "Valental Aspects of Peircean Algebraic Logic", *Computers and Mathematics with Applications* 23/6 (1992), 665–677.

Burch, Robert W., "Peirce's Reduction Thesis", in Nathan Houser, Don D. Roberts, and James Van Evra, ed., *Studies in the Logic of Charles Sanders Peirce* (Indiana University Press, 1997), 234–251.

Burks, Arthur W., "Logic, Learning and Creativity in Evolution", Nathan Houser, Don D. Roberts, and James Van Evra, ed., *Studies in the Logic of Charles Sanders Peirce.* (Indiana University Press, 1997), 497–534.

Calvino, Italo, *Le Cosmicomiche* (Einaudi, 1968 (1965)).

Cameron, Nicole M. et al., "Epigenetic Programming of Phenotypic Variations in Repro-ductive Strategies in the Rat Through Maternal Care", *Journal of Neuroendocrinology* 20/6 (2008), 795–801.

Champagne, Marc, "Kantian Schemata: A Critique Consistent with the Critique", *Philos Investigation* 41 (2018), 436–445.

Chávez, Israel, *The Semiotic Theory of Luis Jorge Prieto* (Tartu University Press, 2022).

Chothia, Cyrus, and Arthur M. Lesk, "The Relation Between the Divergence of Sequence and Structure in Proteins", *The EMBO journal* 5/4 (1986), 823–826.

Cimatti, Felice, *Nel segno del cerchio. L'ontologia semiotica di Giorgio Prodi* (Manifestolibri, 2000).

Cole, Laurence, Krame, Peter (2016).Human Physiology, Biochemistry and Basic Medicine. Elsevier.

Consortium, T. E. P., "The ENCODE Project. An Integrated Encyclopedia of DNA Elements in the Human Genome", *Nature* 489 (2012), 57–74.

Crick, F. 1968 "The origin of the genetic code", *Journal of Molecular Biology*, 38, 3, pp. 367-379, issn: 0022-2836, doi: 10.1016/0022-2836(68)90392-6.

Darwin, Charles, *The Origin of Species* (Wordsworth Editions Limited, 1987).

Dawkins, Richard, *The Selfish Gene* (Oxford University Press, 1976).

Deacon, Terrence W., *Symbolic Species: The Coevolution of Language and Brain* (W.W. Norton & Co, 1997).

Deacon Terrence W., Rączaszek-Leonardi J. *Abandoning the code metaphor is compatible with semiotic process.*, Behav Brain Sci., 2019 Nov 28;42:e224. doi: 10.1017/S0140525X19001419. PMID: 31775915.

De Beaugrande, Robert-Alain, and Wolfgang U. Dressler, *Introduction to Text Linguistics* (Longman, 1981).

Deleuze, Gilles, and Felix Guattari, *A Thousand Plateaus: Capitalism and Schizophrenia* (Bloomsbury Academic: Bloomsbury Revelations, 1987).

Deleuze, Giiles, *The Fold: Leibniz and the Baroque* (Editions de Minuit, 1988).

Dennis, Rutledge M., "Social Darwinism, Scientific Racism, and the Metaphysics of Race", *Journal of Negro Education* 64/3 (1995), 243–252.

Denton, Michael J., Craig J. Marshall, and Michael Legge, "The Protein Folds as Platonic Forms: New Support for the Pre-Darwinian Conception of Evolution by Natural Law", *Journal of Theoretical Biology* 219.3 (2002), 325–342.

Derrida, Jacques, *De la grammatologie* (Editions de Minuit, 1967).

De Saussure, F., *Course in General Linguistics*, trans. by W. Baskin, P. Meisel, and H. Saussy, Columbia University Press (2011).

Eco, Umberto, *Semiotics and the Philosophy of Language. Advances in Semiotics* (Indiana University Press, 1984).

Eco, Umberto, *Kant and the Platypus: Essays on language and cognition* (Harcourt Brace and Company, 1999).

Eco, Umberto, "La soglia e l'infinito", in *Dall'albero al labirinto* (Bompiani, 2007), 463–484.

Faltýnek Dan, Matlach, Vladimír, and Lacková, Ľudmila, "Bases Are Not Letters: On the Analogy Between the Genetic Code and Natural Language by Sequence Analysis", *Biosemiotics* 12/2 (2019), 289–304.

Faltýnek, Dan, Lacková, Ľudmila, "In the Case of Protosemiosis: Indexicality vs. Iconicity of Proteins", *Biosemiotics* 14/1 (2021), 209–226.

Fumagalli, Armando, *Il reale nel linguaggio. Indicalità e realismo nella semiotica di Peirce* (Vita e Pensiero, 1995).

Gare, Arran, "Biosemiosis and Causation: Defending Biosemiotics through Rosen's Theoretical Biology or Integrating Biosemiotics and Anticipatory Systems Theory", *Cosmos and History* 15/1 (2019), 31–90.

Gazzaniga, Michael S., *The Consciousness Instinct* (Farrar Straus and Giroux, 2018).

Gibney, E., and Nolan, C., "Epigenetics and Gene Expression", *Heredity* 105 (2010), 4–13.

Hardwick Charles S., and James Cook, ed., *Semiotic and Significs. The Correspondence between Charles S. Peirce and Victoria Lady Welby* (The Press of Arisbe Associates, 2001).

Harms, William, "Cultural Evolution and the Variable Phenotype", *Biol Philos* 11 (1996), 357–375.

Harris, Roy, *Rationality and the Literate Mind*. Vol. 7. (Routledge, 2009).

Hausman, Carl R., *Charles S. Peirce's Evolutionary Philosophy* (Cambridge University Press, 1993).

Havenel, Jérôme, "Peirce's Topological Concepts", in M. E. Moore, ed., *New Essays on Peirce's Philosophy of Mathematics* (Open Court, 2010).

Hiyama, A., W. Taira, and J. M. Otaki 2012 "Color-Pattern Evolution in Response to Environmental Stress in Butterflies", in *Front. Gene.*

Hjelmslev, Louis, *Principes de grammaire générale* (A.F. Høst. Det., 1928).

Hjelmslev, Louis, *La catégorie des cas: étude de grammaire générale*, Acta Jutlandica, díl 1, Universitetsforlaget, (1935).

Hjelmslev, Louis, *Prolegomena to a Theory of Language*, University of Wisconsin Press, (1963).

Hjelmslev, Louis, "Structure générale des correlations linguistiques", in Louis Hjelmslev, and F. Rastier, *Nouveaux essais* (Presses Universitaires de France, 1985), 25–66.

Houser:Hosid S, Trifonov EN, Bolshoy A., *Sequence periodicity of Escherichia coli is concentrated in intergenic regions*. BMC Mol Biol. 2004 Aug 26;5:14. doi: 10.1186/1471-2199-5-14. PMID: 15333140; PMCID: PMC516772.

Houser, Nathan, "The Intelligible Universe", in Vinicius Romanini, and Elize Fernández, ed., *Peirce and Biosemiotics: A Guess at the Riddle of Life* (Springer Netherlands, 2014), 9–32.

Huneman, Philippe, "Topological Explanations and Robustness in Biological Sciences", *Synthese* 177 (2010), 213–245.

Jablonka, Eva, "Cultural Epigenetics", *The Sociological Review* 64 (2016), 42–60.

Jacob, F., *La logique du vivant: une histoire de l'hérédité*, Bibliothèque des sciences humaines, Gallimard (1970).

Jakobson, Roman, "Linguistics in Relation to Other Sciences", in Roman Jakobson, ed., *Selected Writings Vol 2: Word and Language* (De Gruyter, 1971), 655–696.

Ji, Sungchul, "Isomorphism between Cell and Human Languages: Molecular Biological, Bioinformatic and Linguistic Implications", *Biosystems* 44/1 (1997), 17–39.

Johanson, Arnold, "Modern Topology and Peirce's Theory of the Continuum", *Transactions of the Charles S. Peirce Society*, 37 (2001), 1–12.

Jost E, do O N, Dahl E, Maintz CE, Jousten P, Habets L, Wilop S, Herman JG, Osieka R, Galm O. Epigenetic alterations complement mutation of JAK2 tyrosine kinase in patients with BCR/ABL-negative myeloproliferative disorders. Leukemia. 2007 Mar;21(3):505-10. doi: 10.1038/sj.leu.2404513. Epub 2007 Jan 18. PMID: 17230231.

Kauffman, Stuart. A., *Investigations* (Oxford University Press, 2000).

Ketner, Kenneth L., "Peirce's Most Lucid and Interesting Paper: An Introduction to Cenopythagoreanism", *International Philosophical Quarterly* 26 (1986), 375–392.

Ketner, Kenneth L., "Hartshorne and the Basis of Peirce's Categories", in Robert Kane, and Stephen H. Phillips, ed., *Hartshorne: Process Philosophy and Theology* (State University of New York Press, 1989), 135–150.

Ketner, Kenneth L., and Hilary Putnam, "Introduction: The Consequences of Mathematics", in Charles S. Peirce, Kenneth L. Ketner, ed., *Reasoning and the Logic of Things: The Cambridge Conferences Lectures of 1898.* (Harvard UP, 1992.)

Ketner, Kenneth. L., "Rescuing Science from Scientism", *The Intercollegiate Review* 35/1 (1999), 22–27.

Ketner, Kenneth L., "Charles Sanders Peirce: Interdisciplinary Scientist", in Elize Bisanz, ed., *Charles S. Peirce: The Logic of Interdisciplinarity. The Monist-Series* (De Gruyter, 2009), 35–57.

Ketner, Kenneth L., "Semeiotic", in Elize Bisanz, ed., *Das Bild zwischen Kognition und Kreativität. Interdisziplinäre Zugänge zum bildhaften Denken* (Transcript, 2011a), 375–401.

Ketner, Kenneth L., et al., "Peirce's NonReduction and relational completeness claims in the context of first-order predicate logic", *KODIKAS/CODE: Ars Semeiotica* 34 1/2 (2011b), 3–14.

Ketner, Kenneth L., "Betagraphic: An Alternative Formulation of Predicate Calculus," (Co-Authors: Elize Bisanz, Scott R Cunningham, Clyde Hendrick, Levi Johnson, Kenneth Laine Ketner, Thomas McLaughlin, Michael O'Boyle), accepted for publication in Transactions of the Charles S. Peirce Society , volume 34, (2011c).

Ketner, Kenneth L., "A Survey of Semeiotic as Practice of Reasoning", *Linguistic Frontiers* 6/3 (2023), 1–16.

Ketner, Kenneth L., "Non-self-conscious Interpretants within the Biology of Mind' Festschrift for Claudine Tiercelin". *Klesis Revue Philosophique, Paris* (2024).

Ketner, Kenneth L., *Lowell Lectures of 1903 by Charles S. Peirce* (Peter Lang, 2024).

Kim, Mirang, and Joseph Costello, "DNA Methylation: An Epigenetic Mark of Cellular Memory", *Experimental and Molecular Medicine* 49 (2017), 322.

Lacková, Ľudmila, "A Biosemiotic Encyclopedia: An Encyclopedic Model for Evolution", *Biosemiotics* 11/2 (2018), 307–322.

Lacková, Ľudmila and Zámečník, Lukáš. "Logical Principles of a Topological Explanation: Peirce's iconic logic" Chinese Semiotic Studies, vol. 16, no. 3, 2020, pp. 493-514. https://doi.org/10.1515/css-2020-0027

Lacková, Ľudmila, "Participative Opposition Applied", *Sign Systems Studies* (2021), 261–285.

Lacková, Ľudmila, and Alexander Bolshoy, "Illusions of Linguistics and Illusions of Modern Synthesis: Two Parallel Stories", *Biosemiotics* 4/1 (2021), 115–119.

Lacková, Ľudmila, and Faltýnek, Dan, "The Lower Threshold as a Unifying Principle between Code Biology and Biosemiotics", *Biosystems*, 210 (2021).

Lamarck, Jean-Baptiste, *Zoological Philosophy. An Exposition with Regard to the Natural History of Animals* (Dentu et L'Auteur, 1963 (1809)).

Laubichler, Manfred D. et al., "The Relativity of Biological Function", *Theory in Biosciences* 134 3/4 (2015), 143–147.

Levinthal, Cyrus, "How to Fold Graciously", in P. DeBrunner, J. Tsibris, E. Munck, ed., *Mossbauer Spectroscopy in Biological Systems* (University of Illinois Press, 1969).

Lévy-Bruhl, Lucien, *L'âme Primitive* (Quadrige, 1996).

Liszka, James Jakób, "Peirce's Evolutionary Thought", in T. Thellefsen, B. Sørensen, ed., *Charles Sanders Peirce in His Own Words* (De Gruyter, 2014), 145–152.

Markoš, Anton, *Readers of the Book of Life: Contextualizing Developmental Evolutionary Biology* (Oxford University Press, 2002).

Markoš, Anton, "Přírodní zákony a evoluce", in Jakub Čapek, ed., *Filosofie Henri Bergsona: základní aspekty a problémy* (OIKOYMENH, 2003).

Markoš, Anton, and Josef Kelemen, *Berušky, andělé a stroje* (Dokořán, 2004).

Markoš, Anton, *Náhoda a nutnost: Jacques Monod v zrcadle naší doby: sborník statí* (Pavel Mervart, 2008).

Markoš, Anton, and Jana Švorcova, Recorded Versus Organic Memory: Interaction of Two Worlds as Demonstrated by the Chromatin Dynamics. Biosemiotics 2, 131–149 (2009). https://doi.org/10.1007/s12304-009-9045-5

Markoš, Anton, and Dan Faltýnek, "Language Metaphors of Life", *Biosemiotics* 4 (2011), 171–200.

Markoš, Anton, *Evoluční tápání. Praha: Pavel Mervart* (2016).

Markoš, Anton, Jana Švorcová, and Josef Lhotský, "Living as Languaging: Distributed Knowledge in Living Beings", in Frédéric Vallée-Tourangeau, and Stephen J. Cowley, ed., *Cognition beyond the Brain: Computation, Interactivity and Human Artifice* (Springer, 2013), 193–214.

Markoš, Anton, and Jana Švorcová, *Epigenetic Processes and Evolution of Life* (CRC Press, 2019).

Martinet, André, *Eléments de linguistique générale* (Colin, 1967).

Moczek, Armin P., et al., "The Role of Developmental Plasticity in Evolutionary Innovation", *Proceedings of the Royal Society B: Biological Sciences* 278.1719 (2011), 2705–2713.

Monod, Jacques, *Chance and Necessity* (Alfred A. Knopf, 1971).

Murphey, Murray G., *The Development of Peirce's Philosophy* (Harvard University Press, 1961).

Neumann, John von., *Theory of Self-Reproducing Automata* (University of Illinois Press, 1966).

Noble, Denis, "The Illusions of the Modern Synthesis", *Biosemiotics* 14 (2021), 5–24.

Nöth, Winfried, "24 Peirce's Guess at the Sphinx's Riddle: The Symbol as the Mind's Eyebeam", in Torkild Thellefsen, and Bent Sorensen, ed., *Charles Sanders Peirce in His Own Words: 100 Years of Semiotics, Communication and Cognition. Vol. 14* (De Gruyter, 2014), 153–160.

Otis, Laura, *Organic Memory: History and the Body in the Late Nineteenth and Early Twentieth Centuries* (University of Nebraska Press, 1994).

Owsianková, Hana, Ondřej Kučera, and Dan Faltýnek, "O příbuznosti lingvistiky a biologie", *Studie z aplikované lingvistiky-Studies in Applied Linguistics* 11/1 (2020), 79–93.

Paolucci, Claudio, "Piegature della continuità. Semiotica interpretativa e semiotica generativa", *Versus. Quaderni di studi semiotici* 97 (2004), 111–150.

Paolucci, Claudio, "Lucien Tesnière autore della logica dei relativi. Su alcune insospettate corrispondenze tra Peirce e lo strutturalismo", *E/C* 2006 (2006), 1–16.

Paolucci, Claudio, Strutturalismo e interpretazione, Strumenti Bompiani, Bompiani (2010).

Paolucci, Claudio, "Iconismo primario e gnoseologia semiotica", *Versus* 120 (2015), 135–150.

Pattee, Howard H., "Physical and Functional Conditions for Symbols, Codes, and Languages", *Biosemiotics* 1 (2008), 147–168.

Peirce, Charles S., "The Fixation of Belief", *Popular Science Monthly* 12 (1877), 1–15.

Peirce, Charles S., "Prolegomena to an Apology for Pragmaticism", *The Monist* 16 (1906), 492–546.

Peirce, Charles S., *The Collected Papers of Charles Sanders Peirce*. Arthur W. Burks, ed. (Harvard University Press, 1958).

Peirce, Charles S. *The Essential Peirce, Volume 1: Selected Philosophical Writings (1867–1893). Vol. 1.* (Indiana University Press, 1992).

Peirce, Charles S., *Reasoning and the Logic of Things. The Cambridge Conferences Lectures of 1898*, Kenneth L. Ketner, ed. (Harvard University Press, 1992).

Peirce, Charles S., *The Collected Papers of Charles Sanders Peirce*. Charles Hartshorne, Paul Weiss, ed. (Harvard University Press, 1994).

Peirce, Charles S., *The New Elements of Mathematics*, Vol. I–IV. Carolyn Eisele, ed. (De Gruyter, 1976)

Peirce, Charles S., *Chance, Love, and Logic: Philosophical Essays* (University of Nebraska Press, 1998).

Peirce, Charles S., "The Architecture of Theories", in Elize Bisanz ed., *The Logic of Interdisciplinarity. The Monist Series: Herausgegeben von Elize Bisanz.* Vol. 20. (De Gruyter, 2009) 58–69.

Peirce, Charles S., "Law of Mind", in Elize Bisanz ed., *The Logic of Interdisciplinarity. The Monist Series: Herausgegeben von Elize Bisanz.* Vol. 20. (De Gruyter, 2009), 82–101.

Peirce, Charles S., "The Logic of Interdisciplinarity", in Elize Bisanz ed., *The Logic of Interdisciplinarity. The Monist Series: Herausgegeben von Elize Bisanz.* Vol. 20 (De Gruyter, 2009).

Pelkey, Jamin, "Nonlinear Process in Peirce", *Semiotics* (2012), 77–85.

Peter E Wright, H. Jane Dyson, *Intrinsically unstructured proteins: re-assessing the protein structure-function paradigm*, Journal of Molecular Biology, Volume 293, Issue 2, 1999.

Peter E Wright, H. Jane Dyson, *Intrinsically unstructured proteins and their function*, Nat Rev Mol Cell Biol, Mar 2005, 6(3):197-208. doi: 10.1038/nrm1589. PMID: 15738986.

Piaget, Jean, *Structuralism* (New York: Basic Books, 1970 (1968)).

Pietarinen, Ahti-Veikko, *Signs of Logic: Peircean Themes on the Philosophy of Language, Games, and Communication* (Springer, 2006).

Pigliucci, Massimo, "Genotype–phenotype Mapping and the End of the 'Genes as Blueprint' Metaphor", *Philosophical Transactions. The Royal Society B* 365 (2010), 557–566.

Pietarinen, Ahti-Veikko, "Iconic Logic of Existential Graphs: A Case Study of Commands", in G. Stapleton, J. Howse, J. Lee, ed., *Diagrammatic Representation and Inference. Diagrams 2008. Lecture Notes in Computer Science* 5223 (Springer, 2008), 404–407.

Pietarinen, Ahti-Veikko, "The Peirce–Baldwin Effect and Its Contemporary Significance", *Nordic Conference on Cognitive Semiotics* 2011.

Pietarinen, Ahti-Veikko, and Frederik Stjernfelt, "Peirce and Diagrams: Two Contributors to an Actual Discussion Review Each Other", *Synthese* 192 (2015), 1073–1088.

Prodi, G., L'individuo e la sua firma : biologia e cambiamento antropologico. Bologna : Il Mulino (1989).

Prieto, Luis J., *Pertinence et Pratique* (Éditions de Minuit, 1975).

Reed, Hilary C., et al., "Alternative Splicing Modulates Ubx Protein Function in Drosophila Melanogaster", *Genetics* 184/3 (2010), 745–758.

Rosenberg, Alex and McShea, Daniel, Philosophy of Biology A Contemporary Introduction. Routledge, (2008).

Robertson, M. P. and G. F. Joyce, "The Origins of the RNA World", *Cold Spring Harbor Perspectives in Biology*, 4, 5 (Apr. 2010), pp. 3608-3608, doi: 10.1101/cshperspect.a003608.

Santaella, Braga, L., "Peirce and Biology", *Semiotica* 127 1/4 (1999), 5–21.

Scott, Frances W., *C. S. Peirce's System of Science. Life as a Laboratory* (The Press of Arisbe Associates, 2006).

Sharov, Alexei A., "Functional Information: Towards Synthesis of Biosemiotics and Cybernetics", *Entropy* 12/5 (2010), 1050–1070.

Sharov, Alexei A., and Tommi Vehkavaara, "Protosemiosis: Agency with Reduced Representation Capacity", *Biosemiotics* 8/1 (2015), 103–123.

Shin, Sun-Joo, *The Iconic Logic of Peirce's Graphs* (MIT Press, 2002).

Short, Thomas L., "Interpreting Peirce's Interpretant", *Transactions of the C. S. Peirce Society* XXXII/84 (1996).

Stevenson, Angus, ed., *Oxford Dictionary of English* (Oxford: Oxford University Press, 2010).

Stjernfelt, Frederik, *Diagrammatology. An Investigation on the Borderlines of Phenomenology, Ontology, and Semiotics* (Dordrecht: Springer, 2007).

Stjernfelt, Frederi, *Sheets, diagrams, and realism in Peirce.* (De Gruyter, 2022).

Švorcová, Jana, *Organic memory in embryonic development.* Dissertation, Charles University, Prague (2012).

Švorcová, Jana, and Karel Kleisner, "Evolution by Meaning Attribution: Notes on Biosemiotic Interpretations of Extended Evolutionary Synthesis", *Biosemiotics* 11 (2018), 231–244.

Tessera, Marc, "Is Pre-Darwinian Evolution Plausible?", *Biology direct* 13 (2018), 1–18.

Tollefsbol, Trygve (ed.), *Transgenerational Epigenetics. Volume 13, 2nd Edition* (Elsevier, 2019).

Trifonov, Edward N., "Translation Framing Code and Frame-monitoring Mechanism as Suggested by the Analysis of mRNA and 16S rRNA Nucleotide Sequences", *Journal of Molecular Biology* 194 (1987), 643–652.

Trubetzkoy, Nikolaï S., *Grundzüge der Phonologie. Travaux du Cercle linguistique de Prague* (Kraus Reprint, 1939).

Waddington, Conrad H., *The Strategy of the Genes* (Allen & Unwin, 1957).

Watson Jameson D., and Francis H. Crick, "A Structure for Deoxyribose Nucleic Acid", *Nature* 171 (1953), 737–738.

Weaver, I.an C. G. et al., "Epigenetic Programming by Maternal Behavior", *Nature Neuroscience* 7/8 (2004), 847–854.

Whitney, W. D. ed., *Century Dictionary* (Century Co, 1889–1891).

Zalamea, Fernando, "Peirce's Inversions of the Topological and the Logical: Forgotten Roads for Our Contemporary World", *Rivista di storia della filosofia: LXXII, 3* (2017), 415–434.

Zeman, Joseph J. *The Graphical Logic of C. S. Peirce* (University of Chicago: Department of Philosophy, Ph.D. Dissertation, 1964).

Zeman, Joseph J., "Peirce's Philosophy of Logic", *Transactions of the Charles S. Peirce Society* 22 (1986), 1–22.

Zengiaro, Nicola, *From biosemiotics to physiosemiotics. Towards a speculative semiotics of the inorganic world*, Linguistic Frontiers (2022), Sciendo, vol. 5 no. 3, pp. 37-48. https://doi.org/10.2478/lf-2022-0019

Previous Volumes

Number 1
Studies in Peirce's Semiotic
Institute for Studies in Pragmaticism, 1979

Number 2
Peirce's Conception of God
Donna M. Orange, 1984

Number 3
Peirce's Theory of Scientific Discovery
Richard Tursman, 1987

Number 4
The Semeiosis of Poetic Metaphor
Michael Cabot Haley, 1988

Number 5
Peirce's Philosophy of Religion
Michael L. Raposa, 1989

Number 6
Memorial Appreciations
W. V. O. Quine *et al.*, 1999

Number 7
C. S. Peirce's System of Science
Frances Williams Scott, 2006

Number 8
Semiotic and significs
The Correspondence between
C. S. Peirce and Victoria Lady Welby
Charles S. Hardwick (ed.), 2001

Number 9
The Ralph Gregory Beil
Memorial Volume: Papers in Theoretical Physics
Thomas G. McLaughlin, Ph.D., 2012

Number 10
Applied Interdisciplinary Peirce Studies
Elize Bisanz, 2019

Number 11
On the Logic of Drawing History from
Symbols, Especially from Images
Elize Bisanz / Stephanie Schneider (eds.), 2024.

Number 12
Language of Life
A Peircean Approach to Living Organisms
Ľudmila Lacková, 2025.

www.peterlang.com

www.ingramcontent.com/pod-product-compliance
Lightning Source LLC
Chambersburg PA
CBHW030243100426
42812CB00002B/300